Quieter Pavements Guidance Document

Natural Resource Technical Report NPS/NSNS/NRTR—2013/760

Richard Sohaney[1], Robert O. Rasmussen[1], Paul Donavan[2], Judith L. Rochat[3]

[1]The Transtec Group, Inc.
6111 Balcones Dr
Austin, TX 78731

[2]Illingworth & Rodkin, Inc.
505 Petaluma Blvd South
Petaluma, CA 94952

[3]U.S. Department of Transportation
Research and Innovative Technology Administration
Volpe National Transportation Systems Center
Environmental Measurement and Modeling Division
55 Broadway, RVT-41
Cambridge, MA 02142

June 2013

U.S. Department of the Interior
National Park Service
Natural Resource Stewardship and Science
Fort Collins, Colorado

The National Park Service, Natural Resource Stewardship and Science office in Fort Collins, Colorado, publishes a range of reports that address natural resource topics. These reports are of interest and applicability to a broad audience in the National Park Service and others in natural resource management, including scientists, conservation and environmental constituencies, and the public.

The Natural Resource Technical Report Series is used to disseminate results of scientific studies in the physical, biological, and social sciences for both the advancement of science and the achievement of the National Park Service mission. The series provides contributors with a forum for displaying comprehensive data that are often deleted from journals because of page limitations.

All manuscripts in the series receive the appropriate level of peer review to ensure that the information is scientifically credible, technically accurate, appropriately written for the intended audience, and designed and published in a professional manner.

This report received formal peer review by subject-matter experts who were not directly involved in the collection, analysis, or reporting of the data, and whose background and expertise put them on par technically and scientifically with the authors of the information.

Views, statements, findings, conclusions, recommendations, and data in this report do not necessarily reflect views and policies of the National Park Service, U.S. Department of the Interior. Mention of trade names or commercial products does not constitute endorsement or recommendation for use by the U.S. Government.

This report is available from the Natural Resource Publications Management website (http://www.nature.nps.gov/publications/nrpm/).

Please cite this publication as:

Sohaney, R. R. O. Rasmussen, P. Donavan, and J. L. Rochat. 2013. Quieter pavements guidance document. Natural Resource Technical Report NPS/NSNS/NRTR—2013/760. National Park Service, Fort Collins, Colorado.

NPS 999/121284, June 2013

NOTICE

This document is disseminated under the sponsorship of the Department of Transportation in the interest of information exchange. The United States Government assumes no liability for its contents or use thereof. This report does not constitute a standard, specification, or regulation.

NOTICE

The United States Government does not endorse products or manufacturers. Trade or manufacturers' names appear herein solely because they are considered essential to the objective of this report.

REPORT DOCUMENTATION PAGE		Form Approved OMB No. 0704-0188
Public reporting burden for this collection of information is estimated to average 1 hour per response, including the time for reviewing instructions, searching existing data sources, gathering and maintaining the data needed, and completing and reviewing the collection of information. Send comments regarding this burden estimate or any other aspect of this collection of information, including suggestions for reducing this burden, to Washington Headquarters Services, Directorate for Information Operations and Reports, 1215 Jefferson Davis Highway, Suite 1204, Arlington, VA 22202-4302, and to the Office of Management and Budget, Paperwork Reduction Project (0704-0188), Washington, DC 20503.		
1. AGENCY USE ONLY (Leave blank)	2. REPORT DATE February 2013	3. REPORT TYPE AND DATES COVERED Final Report
4. TITLE AND SUBTITLE Quieter Pavements Guidance Document		5. FUNDING NUMBERS VX82/JT311 VX82/JT312
6. AUTHOR(S) Richard Sohaney[2], Robert O. Rasmussen[2], Paul Donavan[3], Judith L. Rochat[1]		
7. PERFORMING ORGANIZATION NAME(S) AND ADDRESS(ES) (1) U.S. Department of Transportation Research and Innovative Technology Administration John A. Volpe National Transportation Systems Center Environmental Measurement and Modeling Division, RVT-41 Acoustics Facility Cambridge, MA 02142 (2) The Transtec Group, Inc. 6111 Balcones Drive Austin, TX 78731 (3) Illingworth & Rodkin, Inc. 505 Petaluma Blvd. South Petaluma, CA 94952		8. PERFORMING ORGANIZATION REPORT NUMBER DOT-VNTSC-NPS-11-16
9. SPONSORING/MONITORING AGENCY NAME(S) AND ADDRESS(ES) U.S. Department of Interior National Park Service (NPS) Natural Resource Program Center Natural Sounds Program Fort Collins, CO 80525		10. SPONSORING/MONITORING AGENCY REPORT NUMBER NPS/NSNS/NRTR—2013/760
11. SUPPLEMENTARY NOTES NPS program managers: Randy Stanley, Vicki McCusker, Karen Treviño		
12a. DISTRIBUTION/AVAILABILITY STATEMENT		12b. DISTRIBUTION CODE
13. ABSTRACT (Maximum 200 words) This report provides guidance and better practice recommendations to the National Park Service for selecting pavement surfaces to minimize tire-pavement noise. The report contains an overview of common technologies and methods for quieter pavements, descriptions of research and quieter pavement specifications developed by several state agencies, and a directory of state agency noise and materials/pavement engineers. A brief introduction to some of the fundamentals of tire-pavement noise is included in an appendix.		
14. SUBJECT TERMS Quieter pavements, tire/pavement interaction noise, asphalt pavements, concrete pavements, highway traffic noise, vehicle noise		15. NUMBER OF PAGES 68
		16. PRICE CODE
17. SECURITY CLASSIFICATION OF REPORT Unclassified	18. SECURITY CLASSIFICATION OF THIS PAGE Unclassified	19. SECURITY CLASSIFICATION OF ABSTRACT Unclassified	20. LIMITATION OF ABSTRACT

NSN 7540-01-280-5500

Standard Form 298 (Rev. 2-89)
Prescribed by ANSI Std. 239-18
298-102

METRIC/ENGLISH CONVERSION FACTORS

ENGLISH TO METRIC

LENGTH (APPROXIMATE)
- 1 inch (in) = 2.5 centimeters (cm)
- 1 foot (ft) = 30 centimeters (cm)
- 1 yard (yd) = 0.9 meter (m)
- 1 mile (mi) = 1.6 kilometers (km)

AREA (APPROXIMATE)
- 1 square inch (sq in, in^2) = 6.5 square centimeters (cm^2)
- 1 square foot (sq ft, ft^2) = 0.09 square meter (m^2)
- 1 square yard (sq yd, yd^2) = 0.8 square meter (m^2)
- 1 square mile (sq mi, mi^2) = 2.6 square kilometers (km^2)
- 1 acre = 0.4 hectare (he) = 4,000 square meters (m^2)

MASS - WEIGHT (APPROXIMATE)
- 1 ounce (oz) = 28 grams (gm)
- 1 pound (lb) = 0.45 kilogram (kg)
- 1 short ton = 2,000 pounds (lb) = 0.9 tonne (t)

VOLUME (APPROXIMATE)
- 1 teaspoon (tsp) = 5 milliliters (ml)
- 1 tablespoon (tbsp) = 15 milliliters (ml)
- 1 fluid ounce (fl oz) = 30 milliliters (ml)
- 1 cup (c) = 0.24 liter (l)
- 1 pint (pt) = 0.47 liter (l)
- 1 quart (qt) = 0.96 liter (l)
- 1 gallon (gal) = 3.8 liters (l)
- 1 cubic foot (cu ft, ft^3) = 0.03 cubic meter (m^3)
- 1 cubic yard (cu yd, yd^3) = 0.76 cubic meter (m^3)

TEMPERATURE (EXACT)
[(x-32)(5/9)] °F = y °C

METRIC TO ENGLISH

LENGTH (APPROXIMATE)
- 1 millimeter (mm) = 0.04 inch (in)
- 1 centimeter (cm) = 0.4 inch (in)
- 1 meter (m) = 3.3 feet (ft)
- 1 meter (m) = 1.1 yards (yd)
- 1 kilometer (km) = 0.6 mile (mi)

AREA (APPROXIMATE)
- 1 square centimeter (cm^2) = 0.16 square inch (sq in, in^2)
- 1 square meter (m^2) = 1.2 square yards (sq yd, yd^2)
- 1 square kilometer (km^2) = 0.4 square mile (sq mi, mi^2)
- 10,000 square meters (m^2) = 1 hectare (ha) = 2.5 acres

MASS - WEIGHT (APPROXIMATE)
- 1 gram (gm) = 0.036 ounce (oz)
- 1 kilogram (kg) = 2.2 pounds (lb)
- 1 tonne (t) = 1,000 kilograms (kg) = 1.1 short tons

VOLUME (APPROXIMATE)
- 1 milliliter (ml) = 0.03 fluid ounce (fl oz)
- 1 liter (l) = 2.1 pints (pt)
- 1 liter (l) = 1.06 quarts (qt)
- 1 liter (l) = 0.26 gallon (gal)
- 1 cubic meter (m^3) = 36 cubic feet (cu ft, ft^3)
- 1 cubic meter (m^3) = 1.3 cubic yards (cu yd, yd^3)

TEMPERATURE (EXACT)
[(9/5) y + 32] °C = x °F

QUICK INCH - CENTIMETER LENGTH CONVERSION

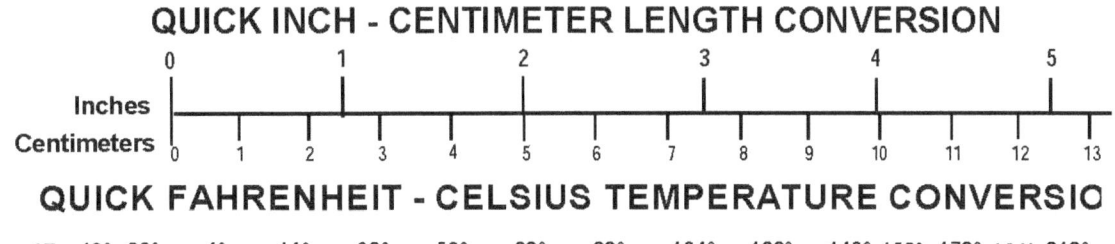

QUICK FAHRENHEIT - CELSIUS TEMPERATURE CONVERSION

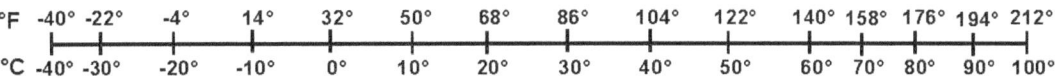

For more exact and or other conversion factors, see NIST Miscellaneous Publication 286, Units of Weights and Measures. Price $2.50 SD Catalog No. C13 10286

ACKNOWLEDGEMENTS

The authors wish to thank the National Park Service Natural Sounds Program, particularly Karen Treviño, Randy Stanley, and Vicki McCusker, for their vision and support of this work.

TABLE OF CONTENTS

Section	Page
ACKNOWLEDGEMENTS	ix
TABLE OF CONTENTS	x
LIST OF FIGURES	xii
LIST OF TABLES	xiii
LIST OF ACRONYMS	xiv
1. INTRODUCTION	1
1.1 Purpose	1
1.2 Contents	1
1.3 Scope	1
2. QUIETER PAVEMENTS REFERENCE	2
2.1 Quieter Pavement Technologies	2
2.1.1 Principles	2
2.1.2 Common Technologies for Flexible (HMA) Pavements	3
2.1.3 Common Technologies for Rigid (Concrete) Pavements	4
2.1.4 Quieter Pavements Summary	5
2.2 State Agency Research and Specifications	10
2.2.1 Arizona Department of Transportation	11
2.2.2 California Department of Transportation	13
2.2.3 Colorado Department of Transportation	16
2.2.4 Florida Department of Transportation	17
2.2.5 Kansas Department of Transportation	17
2.2.6 New Jersey Department of Transportation	19
2.2.7 Ohio Department of Transportation	20
2.2.8 Texas Department of Transportation	21
2.2.9 Virginia Department of Transportation	23
2.2.10 Washington State Department of Transportation	23
2.3 Research Centers and Test Roads	25

TABLE OF CONTENTS (continued)

Section | **Page**

    2.3.1 National Center for Asphalt Technology 25

    2.3.2 National Concrete Pavement Technology Center 25

    2.3.3 MnROAD 26

3. DIRECTORY 29

3.1 State Agency Contacts 30

REFERENCES 39

APPENDIX A. BASICS OF TIRE-PAVEMENT NOISE 44

A.1 Sound vs. Noise 44

A.2 An Analogy 44

A.3 Source-Path-Receiver 44

A.4 Units of Sound: dB 45

A.5 Frequency and A-Weighting 46

A.6 Traffic Noise and Tire-Pavement Noise 47

A.7 Tire-Pavement Noise Generation 48

A.8 How Tire-Pavement Noise is Measured 49

LIST OF FIGURES

Figure **Page**

Figure 1. Porous HMA schematic and photo. 3

Figure 2. Drag (left) and diamond ground (right) concrete pavements. 4

Figure 3. Wider/louder joint design (left) and narrower/quieter joint design (right). 5

Figure 4. Sound pressure amplitudes and the decibel scale for sound level. 45

Figure 5. A-weighting curve. 47

Figure 6. Components of a tire and tread. (Source: Yokohama tires.) 49

Figure 7. Test setup for wayside noise measurements. 50

Figure 8. Photo of microphones in OBSI test configuration. 51

LIST OF TABLES

Table   **Page**

Table 1. Quieter HMA pavement description, elements, and typical noise level range. 6

Table 2. Quieter HMA pavement climatic considerations and regions used. 7

Table 3. Quieter concrete pavement description, elements, and typical noise level range 8

Table 4. Quieter concrete pavement climatic considerations and regions used. 9

Table 5. Typical crossover speeds. 48

LIST OF ACRONYMS

Listed in alphabetical order by acronym

ADOT	Arizona Department of Transportation
ARFC	Asphalt Rubber Friction Course
Caltrans	California Department of Transportation
CDOT	Colorado Department of Transportation
CPB	Controlled Pass-By Method
CP Tech Center	National Concrete Pavement Technology Center
CPX	Close-Proximity Method
CTIM	Continuous-Flow Traffic Time-Integrated Method
DG	Dense-Graded
FDOT	Florida DOT
HMA	Hot-Mix Asphalt
KDOT	Kansas Department of Transportation
MnDOT	Minnesota DOT
MnROAD	MnDOT Pavement Test Track
NCAT	National Center for Asphalt Technology
NJDOT	New Jersey Department of Transportation
OBSI	On-Board Sound Intensity Method
OGAC	Open-Graded Friction Course
OGAC-AR	Open-Graded Friction Course with a Rubber-Modified Binder
OGAC-SBS	Open-Graded Friction Course with a Polymer-Modified Binder
ODOT	Ohio Department of Transportation
QPPP	Quiet Pavement Pilot Program
PEM	Porous European Mix
PFC	Permeable/Porous Friction Course
PCC	Portland Cement Concrete
QPR	Quiet Pavement Research
RHMA	Rubberized Hot-Mix Asphalt
RHMA-G	Gap-Graded Rubberized Hot-Mix Asphalt
RHMA-O	Open-Graded Hot-Mix Asphalt
RHMA-O-HB	Open-Graded High Binder Hot-Mix Asphalt
SIP	Statistical Isolated Pass-By Method
SMA	Stone Matrix Asphalt
SPB	Statistical Pass-By Method
TxDOT	Texas Department of Transportation
UCPRC	University of California Pavement Research Center
USDOT	United States Department of Transportation
VDOT	Virginia Department of Transportation
WSDOT	Washington State Department of Transportation

1. INTRODUCTION

1.1 Purpose

Quieter pavements have been identified as a possible solution to reduce road noise in some National Parks. However, parks looking to reduce road noise need information regarding which quieter pavements might work based on site-specific factors including weather, traffic, and available materials and construction expertise. The purpose of this document is to provide guidance and recommendations on quieter pavements to help park personnel select pavement surfaces that minimize tire-pavement noise.

1.2 Contents

Section 2 starts with an overview of common technologies and methods for quieter pavements. This is followed by a description of research and quieter pavement specifications developed by several state agencies. The agencies selected for inclusion in Section 2 are those with a quiet pavement pilot program (QPPP), a quiet pavement research (QPR) plan, or an otherwise mature investigation of quieter pavement alternatives.

Section 3 contains a directory of state agency noise and materials/pavement engineers.

The appendix contains a brief introduction to some of the fundamentals of tire-pavement noise.

1.3 Scope

This document addresses noise generated from the tire-road interaction on the travel lanes, and not noise generated by rumble strips. Guidance and recommendations on rumble strips is addressed in a separate, companion document.

2. QUIETER PAVEMENTS REFERENCE

The first part of this section presents common techniques associated with quieter pavements. Basic principles that lead to quieter pavements are listed. This is followed by general methods to reduce noise for both flexible (hot-mix asphalt or HMA) pavements and rigid (concrete) pavements. Finally, there are some summary tables for various quieter pavements.

The second part of this section presents quieter pavement specifications developed by some state agencies and a description of the extent of research behind their development.

The third part presents some major pavement research centers and associated test tracks conducting research related to tire-pavement noise.

2.1 Quieter Pavement Technologies

2.1.1 Principles

Pavement research has shown three fundamental principles that lead to reduced tire-pavement noise.

1) **Surface texture** – In general, surface texture should be small, flat (smooth), and negative. Negative texture goes downward into the road surface and away from the tire. Avoid texture that is "positive" with upward going peaks that jab into the tires causing tire vibration and noise.

2) **Porosity** – A porous surface provides channels and opening for air to escape as the tire rolls across. This reduces airflow resistance and other aerodynamic effects and therefore, leads to reduced noise. In addition, surface porosity can increase the acoustic absorption properties of the road surface. Increased absorption helps reduce noise by reducing the amount of noise that propagates away from the tire-pavement interface.

3) **Stiffness** – As a tire rolls on a pavement, the action of the tread coming into contact with the surface is actually a series of impact events. A lower stiffness (softer) surface reduces the strength of the impact and reduces noise.

Quieter pavement technologies invoke one or more of these principles to achieve improved noise reducing performance.

2.1.2 Common Technologies for Flexible (HMA) Pavements

Quieter HMA pavements developed by state agencies use specific gradation requirements and/or polymer or rubber modified binders to achieve their improved noise reducing performance. Common gradations include what are termed "open-graded" and "gap-graded", in which a specified size range of aggregate is excluded from the mix to promote the formation of voids in the pavement. Pavements with open-graded mixtures are often constructed as a surface layer, and are thus commonly cited as an *open-graded friction course* (OGFC) or *permeable/porous friction course* (PFC). Gap-graded HMA mixtures are often termed *stone matrix asphalt* (SMA).

Open gradations achieve a porous surface by providing air voids in the pavement, as shown in Figure 1. The amount of noise reduction is dependent, in part, on the degree of interconnectedness of the voids. Also important to noise reduction is the maximum aggregate size. Most quieter pavement specifications require a nominal maximum aggregate size of 0.5 inches or less. In addition to reduced noise, open-graded pavement surfaces often provide improved friction, hydroplaning resistance, and splash and spray performance.

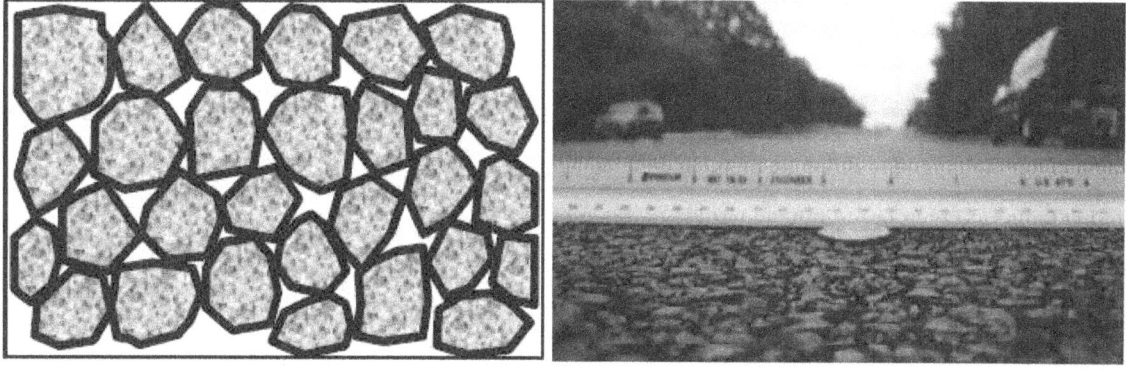

Figure 1. Porous HMA schematic and photo.

Polymer modifiers are commonly used in open and gap graded mixtures in order to improve performance. Sometimes, rubber is added to the mixtures, often in the form of crumb rubber. An HMA layer with rubber is sometimes termed an *asphalt rubber friction course* (ARFC) or *rubberized hot-mix asphalt* (RHMA). In addition to improving performance, the use of scrap tire rubber has the benefit of promoting recycling. More information about crumb rubber and asphalt is available at www.rubberpavements.org.

A bonded wearing course is a gap or open-graded ultra-thin HMA mixture applied over a thick polymer modified asphalt emulsion membrane. Bonded wearing courses are primarily used for maintenance purposes in high traffic areas as a surface treatment over structurally sound pavements.

2.1.3 Common Technologies for Rigid (Concrete) Pavements

Reduced noise from concrete pavements is achieved through improvements in surface texture and joint design. Common textures for quieter concrete surfaces include drag (burlap and artificial turf), longitudinal grooving, and diamond grinding, as shown in Figure 2. Key to producing quieter surface textures is compatibility between the concrete mixture and the texturing process used during construction. The properties of the surface mortar must support formation of the intended texture, and still meet the other functional performance characteristics including smoothness, friction, wear resistance, etc. When diamond grinding is used, the coarse (larger) aggregate is instead exposed and becomes the wearing surface (as opposed to the mortar). For all concrete pavements, it is recommended that, when possible, the mixture contain a hard, durable aggregate that can provide the required frictional characteristics.

Figure 2. Drag (left) and diamond ground (right) concrete pavements.

Joints between slabs can contribute to tire-pavement noise. Aspects of joint design that promote reduced noise include the use of narrow (over wide) joints, and not allowing the joint sealant – if used – to protrude above the pavement surface. Illustrations of louder and quieter joint designs are provided in Figure 3.

Figure 3. Wider/louder joint design (left) and narrower/quieter joint design (right).

2.1.4 Quieter Pavements Summary

This section presents summary tables of the various types of quieter asphalt and concrete pavements. There are two tables each for HMA and concrete. The first table lists descriptions, key characteristics for the quieter pavement, and typical noise levels from the on-board sound intensity (OBSI) evaluation method [AASHTO OBSI]. (The OBSI test method uses microphones positioned very close to the tire-pavement contact patch and measures noise at the source.) The second table lists climatic considerations important during construction, and considerations important for long-term acoustic performance. In the tables, "new" pavements are up to 4 years in age and "aged" pavements are 5 years or older.

Table 1. Quieter HMA pavement description, elements, and typical noise level range.

	Pavement	Description	Elements of a Quieter Pavement	Typical Noise Levels (OBSI, dBA) New	Typical Noise Levels (OBSI, dBA) Aged
Hot-Mix Asphalt	OGFC	Open-graded HMA. Mixture including significant quantity of air voids, typically used as a surface course.	• 15 to 25% porosity. • Nominal maximum aggregate size ≤ 0.375 in (9.5 mm).	96 – 100	98 – 104
Hot-Mix Asphalt	SMA	Gap-graded HMA. Mixture providing stone-on-stone contact and thus including high quality stone.	• Nominal maximum aggregate size ≤ 0.375 in (9.5 mm).	98 – 102	100 – 106
Hot-Mix Asphalt	DG HMA	Dense-graded HMA. Conventional mixtures, most commonly used today.	• Nominal maximum aggregate size ≤ 0.375 in (9.5 mm).	98 – 102	100 – 106
Hot-Mix Asphalt	ARFC	Open-graded HMA with crumb rubber additive. Reduced air voids compared to OGFC, since binder content is higher.	• Nominal maximum aggregate size ≤ 0.375 in (9.5 mm).	96 – 99	98 – 104
Hot-Mix Asphalt	Bonded wearing course	NovaChip® or similar product. Thin wearing course.	• Nominal maximum aggregate size ≤ 0.375 in (9.5 mm).	98 – 102	100 – 104

Table 2. Quieter HMA pavement climatic considerations and regions used.

Pavement		Climatic Considerations		Regions Used
		During Construction	Long Term	
Hot-Mix Asphalt	OGFC	Minimum temperature during construction commonly specified.	• Potential freeze-thaw issues. • Increase in black ice formation possible. • Studded tires and/or chains can lead to premature raveling. • Use of sand for winter maintenance may diminish acoustical benefit by clogging pores.	Typically used in warmer climates found in southern states.
	SMA	No special considerations.	• No special considerations.	Many. Commonly specified to provide additional frictional performance in colder climates.
	DG HMA	No special considerations.	• No special considerations.	All
	ARFC	Minimum temperature during construction commonly specified.	• Potential freeze-thaw issues. • Increase in black ice formation possible. • Studded tires and/or chains can lead to premature raveling.	Typically used in warmer climates found in southern states.
	Bonded wearing course	Minimum temperature during construction commonly specified.	• No special considerations.	Many

Table 3. Quieter concrete pavement description, elements, and typical noise level range.

	Pavement	Description	Elements of a Quieter Pavement	Typical Noise Levels (OBSI, dBA) New	Typical Noise Levels (OBSI, dBA) Aged
Concrete	Diamond ground	Hardened concrete surface ground using diamond grinding head. Final texture has "corduroy" appearance.	• Blades and spacers selected as a function of concrete aggregate type to minimize presence of "fins". • Narrow joints.	98 – 102	100 – 104
Concrete	Drag textured	Wet concrete surface finished by dragging burlap or artificial turf.	• Narrow joints.	99 – 103	101 – 105
Concrete	Longitudinal tined	Wet concrete surface tined (raked) in the longitudinal direction.	• Tine grooves spaced 0.5 or 0.75 in (12.5 or 19 mm), and ≤ 0.125 in (3 mm) deep. • Narrow joints.	100 – 104	101 – 106
Concrete	Transverse tined	Wet concrete surface tined (raked) in the transverse direction.	• Tine grooves spaced ≤ 0.5 in (12.5 mm), and ≤ 0.125 in (3 mm) deep. • Narrow joints.	101 – 105	102 – 108
Concrete	Pervious	Concrete mixture containing significant air voids.	• 20 to 30% porosity. • Maximum aggregate size ≤ 0.375 in (9.5 mm). • Smooth surface (negative texture).	96 – 100	98 – 104

Table 4. Quieter concrete pavement climatic considerations and regions used.

Pavement		Climatic Considerations		Regions Used
		During Construction	Long Term	
Concrete	Diamond ground	No special considerations.	• No special considerations.	Many
	Drag textured	No special considerations.	• No special considerations.	Many
	Longitudinal tined	No special considerations.	• No special considerations.	Many
	Transverse tined	No special considerations.	• No special considerations.	Many
	Pervious	No special considerations.	• Potential freeze-thaw issues. • Increase in black ice formation possible. • Studded tires and/or chains can lead to premature raveling. • Use of sand for winter maintenance may diminish acoustical benefit by clogging pores.	Most. Primarily parking lot and shoulder applications

2.2 State Agency Research and Specifications

Not all states have developed pavements approved specifically for noise reduction. This is because there are at least three major aspects associated with developing a quieter pavement that demand recognition:

1) Understanding variability of tire-pavement noise among sections of the same nominal type of pavement. This variability in noise is due to variability in the construction process, mix production, and other unknown and/or uncontrolled factors. The specification of an approved quieter pavement should recognize this variability, and if possible, encourage the contractor to control the variables within their means.
2) Acoustical durability, where a quieter pavement will change in its ability to provide reduced noise as a pavement wears and ages. Most pavements become louder with age as the surface wears.
3) The balance between a quieter pavement and other requirements for structural and material performance. Quieter pavements that are overlaid must be compatible with the underlying structural course. Furthermore, quieter pavements must survive the climatic and maintenance demands that are expected, including freeze-thaw cycles and mechanical and chemical snow/ice removal.

Many state agencies have initiated quieter pavement research, by conducting literature surveys, measuring the tire-pavement noise of their standard pavements, or via limited evaluations of a pavement that has shown promise as a quieter pavement (possibly based on observations in another state). Fewer state agencies have taken on the more extensive research needed to address the three aspects above to the point of approving a pavement for use in noise reduction.

The following sections present specifications for quieter pavements that are supported by research conducted by the respective state agencies.

It should be noted that a pavement that is researched and approved as quieter by one state agency may not perform in locations outside of that state. This can be due to differences in construction and materials, climatic factors, and maintenance techniques, as indicated in Table 2 and Table 4.

When selecting a quieter pavement developed by another state agency, consideration must be given to be sure the same specification will be successful in your location.

2.2.1 Arizona Department of Transportation

Since the mid 1990s, the Arizona Department of Transportation (ADOT) has been using an asphalt rubber friction course (ARFC) for highway noise reduction. The agency has a long history with asphalt rubber products with first uses in the 1960s. ARFC was first used as an overlay of HMA pavements on the state's interstate system as a rehabilitation strategy to durability and the raveling problems experienced with conventional friction courses. In the early 1990s, it was determined that AR could also produce some noise reduction benefit as well as provide some added durability over conventional HMA pavement designs.

In the mid 1990s, hourly traffic noise L_{eq} values were measured simultaneously before and after transitions from ARFC to ADOT standard uniform transverse tine Portland cement concrete (PCC) at two locations alongside of Interstate 10 near Phoenix [Scofield 2003]. This procedure assured comparable traffic and environmental conditions for the two adjacent pavement types. Microphones were located 25 and 92 ft (7.5 and 28 m) from the nearest edge of the roadway opposite the ARFC and concrete for the first location and 25 and 86 ft (7.5 and 26 m) for the second location. On average, the ARFC was measured to be 4.7 dB lower in level than the concrete.

As ARFC began being used as an overlay of concrete pavements on freeway expansion projects in the Phoenix area, residents near the freeways noticed the reduction in traffic noise. This positive public reaction and results of the research to that point prompted the agency to initiate a project in 2002 to investigate using ARFC as a potential noise mitigation strategy on a wider basis in the Phoenix area. These events led to ADOT developing a Quiet Pavement Pilot Program (QPPP), which was approved by the FHWA in 2003.

As part of the QPPP plan, ARFC was installed on various test sites in and around the Phoenix metro area and Maricopa County. Initial wayside measurements were conducted following the continuous-flow traffic time-integrated method (CTIM) [AASHTO CTIM and Rochat 2009-1].

Microphones located at a distance of 50 ft. (15 m) found noise reductions of approximately 9 to 11 dB when compared to ADOT's standard concrete pavements with similar reductions in tire-pavement noise source levels. Reductions of 7 to 8 dB have been maintained through year three of the study for two of four measurement sites and about 8 dB through year five for the two other sites. Throughout the entire 115-mile (185 km) project area, the average tire-pavement noise source levels have increased about 1.5 dB over the five-year period.

Prior to the QPPP, ADOT also evaluated alternative HMA designs and concrete texturing techniques. In 2000, ADOT built a series of HMA test sections at rural locations on the interstate highway system. These included multiple samples of the same pavement that were randomized in driving order. These test pavements included the ARFC, a non-rubber ACFC, an SMA, a porous ACFC (P-ACFC), and a porous European mix (PEM). All of these pavements had a maximum aggregate of 0.75 in (19 mm) except for the PEM which was 1.25 in (32 mm). OBSI measurements in 2002 indicated that ARFC produced levels about 3 dB lower than the ACFC, 3.5 dB lower than the SMA, and 4 and 5 dB lower than the P-ACFC and PEM, respectively. After six years in 2008, the increase in level for the ARFC was measured to be 2 dB and about equal to the average of the others. The P-ACFC displayed the largest increase of about 4 dB due to a deterioration of performance in lower frequencies, likely due to raveling of the pavement. PEM displayed almost no increase and was now the second quietest pavement. Of the five pavements, the ARFC still maintained the best noise performance after wear, being 2.5 to almost 6 dB quieter than the other pavements while demonstrating an increase in noise of about 0.33 dB/year for 2 to 8 years after construction.

In 2002, ADOT constructed two experimental concrete sections of random transverse tined and longitudinally tined texture to compare to their standard uniform transverse tining. Measurements were conducted using two methods: 1) wayside methods in general conformance with the statistical isolated pass-by (SIP) method [AASHTO SIP], and 2) OBSI. Results for these surfaces indicated that random transverse texture was 2 to 2.5 dB higher in level than the uniform transverse tined while the longitudinally tined surface was 4.5 to 5 dB lower in level. Another concrete section was later diamond ground and found to produce OBSI levels 1 to 1.5 dB lower than the longitudinally tined section, or up 6.5 dB lower than the ADOT standard uniform transverse tine texture. After this research, ADOT adopted the longitudinal tine texture as the new standard texture for concrete in new construction.

ADOT's ARFC is an open-graded HMA with maximum aggregate size of 0.375 in (9.5 mm). There are temperature requirements for when the ARFC can be placed; the surface temperature must be at least 85° F (29.5° C). The full specification for the ARFC can be found in Section 414 of ADOT's specifications for road and bridge construction [ADOT 2008] and at the following Web page.

http://www.azdot.gov/Highways/ConstGrp/contractors/PDF/2008StandardSpecifications.pdf

As a note of caution regarding geographical and climate limitations, the Arizona noise evaluations of the ARFC are limited to the Phoenix metropolitan area and Maricopa County. This is a dry, warm, desert climate. This material has been used in colder areas of the state such as the Flagstaff area where the daily low temperatures are 17° to 19° F (-8 ° to -7° C) in the winter months with highs of 43° to 46° F (6° to 7° C). The ARFC was used in these areas specifically for its winter durability. Its potential as a noise-reducing surface at these colder locations was not explicitly evaluated.

References used for information in this section: ADOT 2006, ADOT Web, and Brown 2008.

2.2.2 California Department of Transportation

The California Department of Transportation (Caltrans) is fully engaged in the area of quieter pavement and has an ongoing, active research program. The state agency has been conducting such research for more than 10 years in partnership with the University of California Pavement Research Center (UCPRC), Illingworth & Rodkin, USDOT Volpe Center, other private parties, and with other states. The agency claims it has the longest running research program in this topic. Caltrans' research has made significant contributions to the field of quieter pavements.

In 1998, Caltrans constructed a pavement research section on rural Interstate 80 with the intent of documenting the initial and long-term acoustic performance of an open-graded HMA overlay on an older existing dense-graded HMA. Based on wayside data, the overlay produced an initial reduction in traffic noise of 6.5 dB. After ten years, a 4.5 dB reduction was still maintained relative to the pre-overlay pavement. Initially, in this research program, test data was obtained using wayside methods (similar to CTIM). The program has transitioned to include testing using the OBSI method. Testing is expected to continue at least through 12 years.

By 2002, Caltrans was using quieter HMA (RHMA-O) specifically for traffic noise reduction in noise-sensitive areas. Reductions of about 6 dB were achieved both in OBSI level and individual vehicle pass-by (SIP) measurements for an overlay of older concrete on a section of Interstate 280 south of San Francisco.

Another similar overlay was applied on a section of Interstate 5 in the Sacramento metropolitan area, and similar reductions of 5 to 7 dB were found using OBSI and wayside measurements [Pommerenck 2009]. The wayside data consisted of traffic noise measured in five-minute L_{eq} time blocks for the northbound and southbound directions at a distance of 15 ft (4.5 m) from the nearest edge of the roadway. Simultaneous measurements were made at the test locations where the overlay was placed and at control locations just beyond the overlay where the existing concrete remained. The difference between the original concrete and the overlay at the test locations was adjusted for traffic conditions in each time block using the data from the control locations. Over a four-year period, the reductions have degraded by less than 1 dB. Caltrans has also used diamond grinding for noise reduction of concrete pavement, achieving reductions of 1 to 3 dB on older, originally longitudinally tined at-grade surfaces and 3 to 10 dB on transversely textured bridge deck surfaces.

In 2002 and 2003, Caltrans constructed HMA and concrete research test sections in northern, rural Los Angeles and Kern Counties. The HMA sections consisted of dense-graded, open-graded, and rubberized designs. Initial comparisons indicated that the dense-graded HMA was 3 to 4 dB higher in level than the opened-graded and rubberized open-graded HMA based on OBSI, controlled and statistical pass-by measurements for light vehicles. For heavy truck statistical pass-bys, these differences were reduced by 1 to 2 dB. Over a 7.5-year period, the open-graded pavements have increased in OBSI level by about 2.5 dB while the dense-graded

HMA has only increased less than 1 dB. The concrete pavement test sections spanned Caltrans' standard longitudinally tined texture, burlap drag texture, diamond grooving, and diamond grinding. The overall range in OBSI level for these surfaces was about 2.5 dB with the longitudinally tined being the highest and one of the diamond grinding sections the lowest. On average, these surfaces have shown about a 0.1 dB/year rate of noise increase from 2003 to 2010.

In November 2006, Caltrans' quiet pavement research (QPR) plan was approved by the FHWA. Based on their research, Caltrans' has released a Pavement Policy Bulletin containing a list of the surfaces approved as quieter pavement strategies [Caltrans 2009]. The bulletin is found at the following Web page.

http://www.dot.ca.gov/hq/esc/Translab/ope/QP-Memo-&-Bulletin-Oct-15-09.pdf

For flexible/composite pavements, an open-graded rubberized HMA (RHMA-O) and an open-graded friction course (OGFC) are approved as overlay or surface treatment. In each case, use 0.5-inch (12.5-mm) maximum aggregate size (or smaller). Alternatives are a gap-graded rubberized HMA (RHMA-G) and dense-graded HMA, again, using 0.5-inch (12.5-mm) maximum aggregate size or smaller.

For rigid pavements, diamond grinding of the pavement surface is approved. For overlays or surface treatment on rigid pavements, an open graded rubberized HMA with high binder (RHMA-O-HB) and a 0.5-inch (12.5-mm) maximum aggregate size is approved.

The policy cautions that thin HMA overlays for noise reducing purposes should not be used in locations where snow tires and chains are allowed during the winter season. This cautionary note is based on Caltrans' general experience with the performance of HMA in the mountainous regions within the state, for example, Donner Pass. It has been found that snow tires and chains can cause excessive wear and early failure through rutting and raveling.

Additional references used for information in this section: Caltrans 2005, Caltrans 2010, Caltrans Web, Lu 2009, and Rochat 2009-2.

2.2.3 Colorado Department of Transportation

The Colorado Department of Transportation (CDOT) has an ongoing quiet pavement research program. It consists of more than 30 test sections spanning a wide range of pavement types, including both HMA and concrete, in locations across the state. The program was initiated in 2003 and the test sections have been evaluated repeatedly on an annual or near annual basis. As a result, the agency is building up a database of tire-pavement noise associated with pavement type and age.

Early in the research, measurements were conducted using the close proximity (CPX) method [ISO/CD 11819-2]. Later, starting in 2006, measurements have been conducted using both OBSI and wayside methods (SIP and CTIM).

Data from CPX testing in the 2004 report showed longitudinally tined concrete to be, on average, 3.8 dB lower in noise than transversely tined concrete [Hanson 2004-1]. Based on this and demonstration of adequate frictional characteristics, CDOT has since adopted longitudinal tining as their preferred method for texturing concrete pavements. The agency's tining specification is contained in section 412 of their specifications for road and bridge construction [CDOT 2005] and found at the following Web page.

> http://www.coloradodot.info/business/designsupport/construction-specifications/2005-construction-specs/2005book

For HMA pavements, CDOT's test sections include SMA, Superpave (dense-graded HMA), and NovaChip® pavements. Early results for HMA revealed the general trend that the coarser the gradation, the higher the noise level [Hanson 2006]. From CPX test results, the 0.5 in (12.5 mm) mixtures (SMA and Superpave) were the quietest. The loudest HMA section was the NovaChip® surface, averaging about 2 dB higher than the SMA and Superpave.

In 2006, the agency started conducting tire-pavement noise measurements using the OBSI method. Field trials of rubberized HMA mixtures have also been included in recent years, and the research is planned to continue until 2011. Interim reports for this more recent testing do not reveal a single pavement type as the quietest. Rather, the HMA and concrete pavements all exhibit a range of overlapping noise levels [Rasmussen 2009].

The agency is not researching the noise performance of open-graded HMA due to reported performance problems using these materials. Open-graded HMA materials have not been shown to be adequately safe in Colorado's extreme winter conditions, and the surface temperatures required when placing rubberized HMA make it restrictive to construction [CDOT Brochure].

2.2.4 Florida Department of Transportation

To date, the Florida Department of Transportation's (FDOT) work has included an assessment of the tire-pavement noise of various pavement types currently in use by the state. Most recently, the agency sponsored the University of Central Florida to collect measurements from 18 test sections throughout the state. The pavement types include permeable open-graded HMA with polymer-modified binders and some with ground rubber, dense-graded HMA with ground rubber, and concrete with diamond grinding and with burlap drag finishes. Noise measurements were conducted using the OBSI and statistical pass-by (SPB) methods.

From the OBSI test results, the dense-graded HMA pavements exhibited 2 to 3 dB lower noise levels than the open-graded HMA [Wayson 2009]. This result is contrary to the research and experience of other states. However, it should be noted that the gradation used in the FDOT open-graded HMA includes larger coarse aggregate, which is likely the cause of the higher noise levels compared to open-graded HMA. For the two concrete pavements studied, the diamond ground surface was about 0.5 dB lower than the burlap drag surface.

FDOT's specifications for dense-graded HMA and concrete pavements are found in the agency's specifications for road and bridge construction [FDOT 2010] and at the following Web page.

http://www.dot.state.fl.us/specificationsoffice/Implemented/SpecBooks/2010BK.shtm

2.2.5 Kansas Department of Transportation

Kansas State Department of Transportation (KDOT) has conducted limited testing of the effects of concrete surface texture on tire-pavement noise. In 2004, highway US-69 a few miles south of Kansas City was re-constructed and widened from two to four lanes. KDOT used this project to construct test sections in the reconstructed northbound lanes as the sites for a surface texture study. This site is in a wet climate region with winter freeze-thaw cycles.

A description of the study and results are in Brennan 2006. KDOT identified four concrete surface textures for evaluation: longitudinal tining, grinding with three different groove widths, 0.110, 0.120, and 0.130 inches (2.8, 3.05, and 3.3 mm), carpet drag, and artificial turf drag. In addition to these four textures, KDOT's study included evaluating two other effects, (i) narrow joints and (ii) the effects of grinding with and without support jacks (which changes the effective wheelbase during grinding). To evaluate the effects of joint width, two test sections of each texture were constructed; one with normal, 0.375 in (9.5 mm) saw cuts and one with narrow, 0.25 in (6.4 mm) saw cuts. To evaluate the effect of the jacks during grinding, the sections with 0.130 (3.3 mm) width grooves were ground with and without jacks. Grinding with jacks had an effective wheelbase of 12 ft (3.7 m), and without jacks an effective wheelbase of 4 ft (1.2 m).

Noise measurements were conducted using the controlled pass-by (CPB) and OBSI methods. For the pass-by testing, two vehicles were used; a passenger car and a 10 wheel dump truck. Results for the truck were confounded by engine noise because some of the pavement sections were not on a level grade. The two sections with slight uphill grade were loudest and the one section with downhill grade was the quietest. The grade affected the amount of engine noise on the test sections to the extent that grade became a larger effect on the pass-by noise than the surface texture.

For the passenger car results, the drag, the 0.110 in (2.8 mm) grooved, and the 0.120 in (3.05 mm) grooved textures were quieter than the wide 0.130 in (3.3 mm) grooved and longitudinal tined. The longitudinal tined was consistently the loudest. In the pass-by test, the 0.130 in (3.30 mm) ground section was 1.1 dB higher and the longitudinal tined section was 2.3 dB higher than the levels for the quietest sections. In the OBSI test, the 0.130 in (3.30 mm) ground section was 1.9 dB higher and the longitudinal tined section was 4.2 dB higher than the levels for the quietest sections.

Results for the different width joints were mixed. In the pass-by test results, the sections with narrow joints were generally quieter. The sections with narrow joints ranged from 0.2 to 2.2 dB quieter than sections with wide joints. The OBSI test results were not consistent. For most sections, the narrow joints were about 1 dB quieter. However, for the longitudinal tined texture, the narrow joints were 2.1 dB louder.

Results for the effect of effective wheelbase during grinding were not significant. The test sections with varying wheelbases were within 1 dB or less and inconsistent between pass-by and OBSI test methods.

All these results are from testing conducted in 2005, shortly after the construction was complete. Therefore, results presented herein represent the pavement in new condition. These same sections were re-evaluated over the next few years, but the results are not available as of publication of this document. Therefore, at this time, there is no data demonstrating the longevity of the acoustic performance of these concrete surface textures in Kansas.

Kansas' specifications for concrete pavements are in section 500 of their specifications for road and bridge construction [KDOT 2007] and at the following Web page.

http://www.ksdot.org/burConsMain/specprov/2007SSDefault.asp

2.2.6 New Jersey Department of Transportation

The New Jersey Department of Transportation (NJDOT) sponsored research on tire-pavement noise in 2002 to 2003. A total of 42 pavement sections were selected throughout the state for noise evaluation. HMA pavements included open-graded friction course, dense-graded HMA, stone matrix asphalt, NovaChip®, and micro-surfaced pavements. Concrete pavements included diamond ground, transverse grooved, transverse tined, and broom drag finished. The research and testing was carried out by Rutgers University and National Center for Asphalt Technology (NCAT) using the close proximity (CPX) method.

For the HMA pavements, NJDOT found their OGFC with crumb rubber and MOGFC-1 to be the quietest [Bennet 2004]. These pavements ranged from 4.8 to 2.0 dB less than the loudest HMA pavements, which were a 0.75 in (19 mm) Superpave and a 0.5 in (12.5 mm) SMA.

For the concrete pavements, the diamond ground was quietest, averaging 7.6 dB less than the state's loudest pavement which was transverse tined.

NJDOT's pavement specifications are in their specifications for road and bridge construction [NJDOT 2007] and at the following Web page.

http://www.state.nj.us/transportation/eng/specs/2007/Division.shtml

Specifications for the OGFC and MOGFC-1 mixes are in Section 902.03. Specifications for diamond grinding of concrete surfaces are in Section 405.03.04.

2.2.7 Ohio Department of Transportation

The Ohio Department of Transportation (ODOT) has sponsored some research in the area of longitudinal versus transverse texturing on concrete pavements. This work was initiated because of a poor experience on a section of I-76 near Akron, Ohio. The pavement surface prior to reconstruction was HMA. This section was reconstructed with concrete using the state's standard practice of randomly spaced transverse tining. After reconstruction, there were complaints of excessive traffic noise, and so the agency planned to change the concrete surface texture by diamond grinding. Before changing, the agency planned a research project to quantify the noise reduction due to the re-texturing.

The research was carried out by the Ohio Research Institute for Transportation and the Environment at Ohio University [ODOT 2005-2]. Noise measurements were done at five sites using wayside methods with microphones positioned at 25 and 50 ft (7.5 and 15 m) from the centerline of the outside lane. Data were obtained before and after the texture change. The average noise reduction after diamond grinding was 3.5 dB at the 25-ft (7.5-m) microphone, and 3.1 dB at the 50-ft (15-m) microphone.

Based on the research, the agency has added a special provision for longitudinal texturing (via fresh concrete tining). From section 451.09 of the agency's construction and materials specifications [ODOT 2010-1]:

If longitudinal tining is authorized the tine spacing will be a uniform 3/4 inches wide (19 mm), 1/8 inch deep (3 mm) and 1/8 inch wide (3 mm). Do not tine within 3 inches (75 mm) of pavement edges or longitudinal joints. Only use equipment that will tine the full width of the pavement in one operation and uses stringline controls for line and grade to assure straight tining texture.

As an option for HMA pavements, section 409.2 of the agency's pavement design manual [ODOT 2010-2] lists item 803 rubberized open-graded friction course as a special use item for reduced tire pavement noise. Specifications for the pavement appear in a supplemental specification [ODOT 2005-1] and at the following Web page.

http://www.dot.state.oh.us/Divisions/ConstructionMgt/Specs%20and%20Notes%20for%202005/803_04152005%20for%202005.PDF

The design manual cautions that ODOT has experienced difficulty with snow and ice removal from OGFC. In addition, the OGFC does not add structural strength, and must therefore be placed on a pavement with adequate structural capacity. Section 803.05 of the supplemental specification requires the surface temperature to be at least 55 °F (13 °C) and rising at the time of placement.

2.2.8 Texas Department of Transportation

Texas Department of Transportation (TxDOT) has sponsored extensive research on tire-pavement noise associated with use of permeable friction courses (PFC). A PFC is open-graded HMA with typically 18 percent or more air void content. It is sometimes referred to as the "New Generation Open-Graded Friction Course". The pavement is usually selected for its improved safety in wet conditions, reduced splash and spray, and increased resistance to hydroplaning.

Another characteristic of PFC is that it has demonstrated reduced tire-pavement noise. The agency observed that the noise reduction was reported by the driving public. This prompted TxDOT to initiate a five-year study on the tire-pavement noise associated with the use of PFC in Texas [Trevino 2006]. The main objectives of the study are to measure the effects of aging on the acoustical performance of the pavements, and to provide data for possibly using pavement type as part of the FHWA Traffic Noise Model® (TNM®) [Anderson 2004, Fleming 1996, Lee 1996, AASHTO SIP], and thus to determine the need for noise mitigation.

The agency developed an in-state research and data acquisition plan. Their test plan focused on three factors: pavement type, age, and climate. HMA pavement types include plant mix seal, PFC, and dense-graded HMA. Some concrete pavements were tested as well. For pavement age, the agency considered five years of service as defining the boundary between old and new pavements. For climate, the agency identified four climate zones within the state with the different combinations of moisture (wet and dry) and temperature (freeze and no freeze). Overall, more than 30 test sections have been studied in Texas.

Testing has been conducted by both the state agency and by the Center for Transportation Research at the University of Texas at Austin. Tire-pavement noise was measured using the OBSI method, as well as wayside using the statistical pass-by method. [Trevino 2009-1]

Results of TxDOT's research program show that tire-pavement noise from the PFC is almost 3 dB less than other pavements (such as dense-graded HMA and concrete pavement) based on measurements using the OBSI method [Trevino 2009-2]. PFC grows slightly louder with age but remains quieter than Texas' conventional pavements over their life. The increase with age is fastest immediately after construction, but then levels off.

The specifications for the Texas PFC is item 342 of the agency's specifications manual [TxDOT 2004] and is accessible at the following Web page.

ftp://ftp.dot.state.tx.us/pub/txdot-info/cmd/cserve/specs/2004/standard/s342.pdf

2.2.9 Virginia Department of Transportation

Virginia Department of Transportation (VDOT) has developed a specification for an open-graded surface course (also referred to as a porous friction course, or PFC) that is optimized for key functional performance characteristics including ride quality, skid resistance, and tire-pavement noise. The PFC is suitable for situations where lower noise is wanted, but not suitable for providing structural strength. As such, it should be applied as a surface course on an underlying structural course.

The pavement is a result of a demonstration project conducted by utilizing this functionally optimized pavement on an 8000+ ft (2425+ m) test section of a four lane, divided, primary route. Traffic on the test section was relatively high volume with typical mix of passenger vehicles, trucks, and busses traveling at speeds in excess of 35 mph. Tire-pavement noise measurements were conducted for the project using the OBSI method.

Preliminary results show a 6 dB improvement in tire-pavement noise between the road with the original dense-graded HMA pavement and with the PFC pavement in new condition [McGhee 2009 and McGhee 2010]. Results also showed 2 to 3 dB less tire-pavement noise with the PFC pavement compared to other new, conventional pavements, including both dense-graded HMA and stone matrix asphalt.

A special provisional specification for the optimized PFC is contained in Appendix B of McGhee 2009 and at the following Web page.

http://www.virginiadot.org/vtrc/main/online_reports/pdf/09-r20.pdf

2.2.10 Washington State Department of Transportation

Washington State Department of Transportation (WSDOT) has an ongoing research plan investigating two open-graded HMA designs for use as quieter pavements. One design has a rubber modified binder (OGFC-AR), and the other has a polymer modified binder (OGFC-SBS). Both have 0.375-in (9.5-mm) maximum aggregate size, and are modeled after the OGFC (ARFC) used in Arizona.

The test sections are on highways with high volume traffic in and around the Seattle metro area. The first test sections were placed in 2006 on Interstate 5 in Lynnwood, and were followed in 2007 by a second set of test sections on State Route 520 in Medina. The agency's data acquisition plan calls for noise measurements using the OBSI method on a monthly basis, weather permitting. This is one of the highest frequency measurement intervals of all the states, and helps to clearly establish trends due to aging (in addition to seasonal variations in the measurement).

Results from WSDOT's research plan show the OGFC pavements initially are quieter than the state's standard HMA in new condition by 2 to 4 dB. However, after 3 years, the noise levels from the pavements increased to the same levels as the standard HMA, and thus they lost the lower noise advantage.

In addition, the OGFC pavements failed structurally. By 2009, the test sections exhibited excess rutting due to raveling indicating significantly reduced service life. The raveling is due to winter conditions in the state of Washington: frequent precipitation, freeze-thaw cycles, studded tires, snow chains, and snowplows.

WSDOT's research continues to monitor the performance of the OGFC test sections, and recently added test sections of concrete pavement to investigate quieter textures.

For reference, WSDOT's special provisional specification for the OGFC is a modification to the specifications in Section 5-04 of the agency's specifications manual [WSDOT 2010] and accessible at the following Web page.

http://www.wsdot.wa.gov/publications/manuals/fulltext/M41-10/SS2010.pdf

Additional references used for information in this section: WSDOT 2005, WSDOT 2009, WSDOT Web, and Sexton 2010.

2.3 Research Centers and Test Roads

This section lists some major pavement research centers involved with tire-pavement noise.

2.3.1 National Center for Asphalt Technology

The National Center for Asphalt Technology (NCAT), located at Auburn University, Alabama, is a research facility dedicated to hot-mix asphalt. NCAT has a 1.7-mile (2.72-km) test track with more than 40 instrumented 200-ft (61-m) sections for asphalt testing. A fleet of heavy tractor-trailers operates around the track to apply an accelerated loading. In addition to the test track, NCAT has a research laboratory for testing asphalt binders and mixtures, and equipment for testing pavements in the field.

NCAT has been previously involved with tire-pavement noise measurements and research. They constructed a close-proximity (CPX) noise trailer, which is a sound pressure based, tire-pavement noise-measuring device. (This trailer was a common method of measuring tire-pavement noise before the development of the on-board sound intensity method.) Several state agencies had sponsored NCAT to survey the tire-pavement noise of their roadways. In addition, low noise HMA pavements have been installed and studied on the NCAT test track.

Research at NCAT has shown that open-graded friction course mixtures (porous friction courses) have significantly reduced tire-pavement noise as well as other improved surface characteristics such as skid resistance and splash & spray.

References used for information in this section: Hanson 2004-2, Fortier Smit 2007, and Fortier Smit 2008.

2.3.2 National Concrete Pavement Technology Center

The National Concrete Pavement Technology Center (CP Tech Center), located at Iowa State University, is dedicated to concrete pavement research and technology transfer. One of the major research initiatives ongoing at the CP Tech Center is the Concrete Pavement Surface Characteristics Program, which is focusing on tire-pavement noise as well as texture, friction, splash/spray, rolling resistance, reflectivity/illuminance, and smoothness.

CP Tech Center research has produced numerous reports and "Technical Briefs" on the topic of tire-pavement noise for concrete roadways. A major finding from their studies is that more significant than designing new, innovative surface textures is maintaining tighter tolerance and control of standard surface textures (drag, tine, grinding) during construction [Rasmussen 2008].

2.3.3 MnROAD

MnROAD is a road test and research facility specifically for cold climate regions. It is located about 40 miles northwest of the Twin Cities, near Albertville, MN. Although it is closely associated with the Minnesota Department of Transportation (MnDOT) Office of Materials and Road Research, the facility is used by researchers from all over the country and the world. It is recognized as the world's largest and most comprehensive facility for cold weather road research. The research facility has two test tracks. One is a closed loop track on which a multi-axle truck operates as a controlled loading. The other is a section of Interstate 94 which carries normal highway traffic.

Currently, there are three research projects ongoing at MnROAD that are related to pavement surface characteristics, including tire-pavement noise [MnROAD 2009-2]:

1) Concrete Pavement Surface Characteristics, Rehabilitation
2) Concrete Pavement Surface Characteristics, New Construction
3) HMA Pavement Surface Characteristics

The projects include the following surfaces:
- Dense graded HMA
- NovaChip®
- 3/16 in (4.75 mm) HMA
- Porous asphalt
- Longitudinal and transverse tined concrete
- Random transverse tined concrete
- Conventional and innovative diamond grind concrete

- Longitudinal and transverse broom drag concrete
- Artificial turf drag concrete
- Pervious concrete

The projects plan to measure the surfaces over a five-year period to evaluate their noise performance as the pavements age. They started in 2007, thus they are not complete yet. However, some interim results for the concrete grind pavements have been published [Izevbekhai 2007]. When new, the innovative grind concrete was 4.5 dB lower than conventional grind as measured using the OBSI method. Over a two-year period, the innovative grind remained consistently quieter than the conventional grind, although the gap lessened to about 2.5 dB.

An additional reference used for information in this section is MnROAD 2009-1.

3. DIRECTORY

This section presents quieter pavement resources in the form of a directory of state highway agency contacts.

The technology of quieter pavements spans multiple disciplines: acoustics/noise control and pavement materials. Each state highway agency has its own unique organization, but, generally, the people with noise expertise are organized under the agency's environmental division. People with pavement and materials expertise are organized under the agency's materials, pavement, and/or roadway construction divisions.

Section 3.1 is a table listing contact information for a noise engineer and materials/pavement engineer for each state agency (including the District of Columbia and territory of Puerto Rico).

Other Web based resources that can be referenced for the most current information include:
- The Transportation Research Board Committee on Transportation-Related Noise and Vibration Web site includes a list of state agency noise contacts.
    http://www.adc40.org/adc40membershiplist.pdf

- The U.S. Department of Transportation Federal Highway Administration (FHWA) Web site contains a Pavement Contacts page includes state agencies pavement contacts.
    http://www.fhwa.dot.gov/pavement/pavecont.cfm

- The state department of transportation (DOT) Web sites. Most state agencies have Web pages for their environmental, materials, and pavement divisions. States active in the area of quieter pavement usually have project information on their Web site. Also online are the state highway agency's materials and construction standards, which can contain relevant pavement specifications or special provisions.

3.1 State Agency Contacts

Local Agency	Noise Engineer Contact	Materials / Pavement Engineer Contact
Alabama	Daniel Turman Environmental Technical Section Alabama Department of Transportation 1409 Coliseum Blvd. Montgomery, Alabama 36130-3050 Phone: 334-242-6828 Email: turmand@dot.state.al.us	Larry Lockett Materials Engineer Alabama Dept of Transportation 1409 Coliseum Boulevard Montgomery, Alabama 36110 Phone: 334-206-2201
Alaska	Ben White Alaska Department of Transportation 3132 Channel Drive Juneau, Alaska 99801-7898 Phone: 907-465-6961 Email: ben.white@alaska.gov	Steve Saboundjian State Pavement Engineer Alaska Dept. of Transportation and Public Facilities (ADOT&PF) Statewide Materials 5800 East Tudor Road Anchorage, AK 99507-1286 Phone: 907-269-6214
Arizona	Fred Garcia Noise/Air Team Environmental Planning Section Arizona Department of Transportation 205 South 17th Avenue Room 213 Mail Drop 619E Phoenix, Arizona 85007-3212 Phone: 602-712-8635 Email: fgarcia@azdot.gov	James Delton Asst State Engineer for Materials Arizona Dept of Transportation 4000 N. Central Ave Suite 1500 Phoenix, Arizona 85012 Phone: 602-712-8094
Arkansas	Brenda Price Environmental Division Arkansas Department of Transportation Post Office Box 2261 Little Rock, Arkansas 72203 Phone: 501-569-2521 Email: brenda.price@arkansashighways.com	Michael Benson Materials Engineer, Materials Division Head Materials Division Arkansas State Highway and Transportation Department 11301 West Baseline Little Rock, Arkansas 72209 Phone: 501-569-2185
California	Bruce Rymer Caltrans Division of Environmental Analysis P.O. Box 942874 1120 N Street, M.S. 27 Sacramento, California 94274-0001 Phone: 916-653-6073 Email: bruce_rymer@dot.ca.gov	Bill Farnbach Caltrans P.O. Box 942874 Sacramento, California 94274-0001 Phone: 916-274-6188

Local Agency	Noise Engineer Contact	Materials / Pavement Engineer Contact
Colorado	Jill T. Schlaefer Colorado Department of Transportation 4201 E. Arkansas Ave., Shumate Bldg. Denver, Colorado 80222 Phone: 303-757-9016 Email: jill.schlaefer@dot.state.co.us	Jay Goldbaum Pavement Design Manager Colorado Dept of Transportation 4670 Holly Street, Unit A Denver, Colorado 80216-6408 Phone: 303-398-6561
Connecticut	Desmond P. Dickey Office of Environmental Planning Connecticut Department of Transportation Post Office Box 317546 2800 Berlin Turnpike Newington, Connecticut 06131-7546 Phone: 860-594-2945 Email: d.paul.dickey@po.state.ct.us	Wayne Blair Asst. Director, Div. of Materials & Testing Connecticut Department of Transportation 280 West Street Rocky Hill, Connecticut 06067 Phone: 860-258-0312
Delaware	Edwin Kuipers Delaware Department of Transportation Post Office Box 778 Dover, Delaware 19903 Phone: 302-760-2515 Email: ekuipers@mail.dot.state.de.us	Jim Pappas Materials and Research Engineer Delaware Dept of Transportation P.O. Box 778 Dover, Delaware 19903 Phone: 302-760-2400
District of Columbia	Maurice Keys District of Columbia Dept. of Public Works 2000 14th Street, NW Washington, DC 20009 Phone: 202-671-2740 Email: maurice.keys@dc.gov	Wasi Khan Materials Engineer District Department of Transportation 2000 14th Street, N.W. Washington, District of Columbia. 20009 Phone: 202-672-2540
Florida	Mariano Berrios Florida Department of Transportation Environmental Management Office 605 Suwannee Street, MS-37 Tallahassee, Florida 32399-0450 Phone: 850-414-5250 Email: mariano.berrios@dot.state.fl.us	Bouzid Choubane State Pavement Material Systems Engineer State Materials Office 5007 NE 39th Avenue Gainesville, FL 32609 Phone: 352-955-6302
Georgia	Keisha Jackson Georgia Department of Transportation 16th Floor, One Georgia Center 600 W. Peachtree St., NW Atlanta, GA 30308 Phone: 404-631-1160 Email: keijackson@dot.ga.gov	Georgene Geary State Materials and Research Engineer Georgia DOT 15 Kennedy Drive, Forest Park, GA 30297 Phone: 404-608-4700

Local Agency	Noise Engineer Contact	Materials / Pavement Engineer Contact
Hawaii	Steve Ege Hawaii Department of Transportation Highway Division Materials, Testing & Research Branch 2530 Likelike Highway Honolulu, Hawaii 96819 Phone: 808-832-3405 Email: steve.ege@hawaii.gov	Herbert Chu Materials/Research Engineer Hawaii DOT Highway Div. 869 Punchbowl Street Honolulu, Hawaii 96813-5097 Phone: 808-832-3405
Idaho	Roy Jost Idaho Department of Transportation Environmental Section 3311 West State Street Boise, Idaho 83707-1129 Phone: 208-334-8477 Email: roy.jost@itd.idaho.gov	Jeff Miles Materials Engineer Idaho Transportation Dept. P.O. Box 7129 Boise, Idaho 83707 Phone: 208-334-8440
Illinois	Walt Zyznieuski Illinois Department of Transportation 2300 South Dirksen Parkway, Room 330 Springfield, Illinois 62764 Phone: 217-785-4181 Email: Walter.zyznieuski@illinois.gov	David Lippert Materials Engineer Illinois Dept of Transportation 2300 South Dirksen Parkway Springfield, Illinois 62703 Phone: 217-782-7200
Indiana	Ron Bales Office of Environmental Services Indiana Department of Transportation 100 N. Senate Avenue, Room N642 Indianapolis, Indiana 46204-2249 Phone: 317-234-4916 Email: rbales@indot.in.gov	Tommy Nantung Pavement Research Engineer Indiana Dept. of Transportation 100 N. Senate Avenue, N808 Indianapolis, Indiana 46204 Phone: 765-463-1521 x248
Iowa	Charles Bernhard Office of Location and Environment Iowa Department of Transportation 800 Lincoln Way Ames, Iowa 50010 Phone: 515-239-1410 Email: Charles.Bernhard@dot.iowa.gov	Chris Brakke Pavement Engineer Iowa Dept. of Transportation 800 Lincoln Way Ames, Iowa 50010 Phone: 515-239-1882
Kansas	Michael Fletcher Environmental Services Section Kansas Department of Transportation Docking State Office Building 915 SW Harrison Topeka, Kansas 66612-1568 Phone: 785-296-3726 Email: Fletcher@ksdot.org	Andy Gisi Kansas Dept. of Transportation Dwight D. Eisenhower State Office Building 700 SW Harrison Street Topeka, Kansas 66603-3745 Phone: 785 296-2231

Local Agency	Noise Engineer Contact	Materials / Pavement Engineer Contact
Kentucky	Tom Koos Kentucky Transportation Cabinet Division of Environmental Analysis 200 Mero Street Frankfort, Kentucky 40622-1994 Phone: 502-564-7250 Email: Tom.Koos@ky.gov	Wesley Glass Materials Engineer Kentucky Transportation Cabinet 200 Mero Street Frankfort, Kentucky 40622 Phone: 502-564-3160
Louisiana	Noel Ardoin Louisiana Department of Transportation and Development Post Office Box 94245 Baton Rouge, Louisiana 70804-9245 Phone: 225-242-4501 Email: Noel.Ardoin@la.gov	Jeffrey Lambert Pavement Engineer Louisiana Department of Transportation P.O. Box 94245 Baton Rouge, Louisiana 70804-9245 Phone: 225-379-1937
Maine	Nathan Howard Maine Department of Transportation Bureau of Transportation System Planning 16 State House Station Augusta, Maine 04333-0016 Phone: 207-624-3310 Email: Nathan.Howard@maine.gov	Bruce Yeaton Materials Engineer Maine Dept. of Transportation 16 State House Station Augusta, Maine 04333-0016 Phone: 207-287-3482
Maryland	Ken Polcak Maryland State Highway Administration Office of Planning and Preliminary Engineering 707 N. Calvert Street C-503 Baltimore, Maryland 21202 Phone: 410-545-8601 Email: kpolcak@sha.state.md.us	Geoffrey Hall Pavement Division Chief Maryland State Highway Administration 7450 Traffic Drive Hanover, Maryland 21076 Phone: 443-572-5067
Massachusetts	Ryan McNeill Massachusetts Highway Department 10 Park Plaza Boston, Massachusetts 02116-3973 Phone: 617-973-7446 Email: E.McNeill@mhd.state.ma.us	Clement Fung Materials Engineer Massachusetts Highway Dept. 10 Park Plaza Boston, Massachusetts 02116-3973 Phone: 617-973-8441
Michigan	Tom Hanf Michigan Department of Transportation Environmental Analysis Unit Post Office Box 30050 Lansing, Michigan 48909 Phone: 517-241-2445 Email: hanft@michigan.gov	John Staton Engineer of Materials Michigan Dept of Transportation 8885 Ricks Road Lansing, Michigan 48909 Phone: 517-322-5701

Local Agency	Noise Engineer Contact	Materials / Pavement Engineer Contact
Minnesota	Melvin Roseen Minnesota Department of Transportation Noise Analysis Unit 6000 South Minnehaha Avenue Saint Paul, Minnesota 55111 Phone: 651-366-5808 Email: melvin.roseen@state.mn.us	Curt Turgeon Pavement Engineer Minnesota Dept of Transportation 1400 Gervais Avenue Maplewood, Minnesota 55109 Phone: 651-779-5535
Mississippi	Elton D. Holloway Planning Division Mississippi Department of Transportation Post Office Box 1850 Jackson, Mississippi 39215-1850 Phone: 601-359-7685 Email: eholloway@mdot.state.ms.us	Jimmy Hammons Pavement Design Engineer Mississippi Department of Transportation 401 North West Street Jackson, Mississippi 39201 Phone: 601-359-7250
Missouri	Rob Meade Missouri Dept. of Transportation Design-Environmental P.O. Box 270 Jefferson City, Missouri 65102 Phone: 573-526-6677 Email: Robert.Meade@modot.mo.gov	Jay Bledsoe System Analysis Engineer Missouri Department of Transportation P.O. Box 270 Jefferson City, Missouri 65102 Phone: 573-751-3634
Montana	Cora G. Helm Montana Department of Transportation Environmental Services Post Office Box 201001 Helena, Montana 59620-1001 Phone: 406-444-7659 Email: cohelm@mt.gov	Matt Strizich Materials Engineer Montana Department of Transportation 2701 Prospect Avenue Helena, Montana 59620 Phone: 406-444-6297
Nebraska	Mark Ottemann Project Development Division Nebraska Department of Roads Post Office Box 94759 Lincoln, Nebraska 68509-4759 Phone: 402-479-4684 Email: mark.ottemann@nebraska.gov	Moe Jamshidi Materials Engineer Nebraska Department of Roads P.O. Box 94759 1500 Nebraska Highway 2 Lincoln, Nebraska 68509-4759 Phone: 402-479-4750
Nevada	Daniel Harms Environmental Services Division Nevada Department of Transportation 1263 S. Stewart Street Carson City, Nevada 89712 Phone: 775-888-7685 Email: dharms@dot.state.nv.us	Tie He Research Engineer Nevada Dept of Transportation 1263 S. Stewart Street Carlson City, Nevada 89712 Phone: 775-888-7220

Local Agency	Noise Engineer Contact	Materials / Pavement Engineer Contact
New Hampshire	Jonathan Evans Bureau of Environment Room New Hampshire Department of Transportation Post Office Box 483 Concord, New Hampshire 03302-0483 Phone: 603-271-4048 Email: JEvans@dot.state.nh.us	Denis Boisvert Chief of Materials Technology New Hampshire Department of Transportation Bureau of Materials & Research PO Box 483, 5 Hazen Drive Concord, New Hampshire 03302-0483 Phone: 603-271-1545
New Jersey	Lane Liou New Jersey Department of Transportation P.O. Box 600 Trenton, New Jersey 08625 Phone: 609-530-5503 Email: @dot.state.nj.us (call to get email)	Robert Sauber Materials Engineer New Jersey Department of Transportation 1035 Parkway Ave, PO Box 600 Trenton, New Jersey 08625-0600 Phone: 609-530-3755
New Mexico	Jeff Fredine Environmental Section, Room 213 New Mexico State Highway & Transportation Department Post Office Box 1149 Santa Fe, New Mexico 87504-1149 Phone: 505-827-5681 Email: jeffrey.fredine@state.nm.us	Jeffrey Mann Pavement Design New Mexico Department of Transportation Room 224 1120 Cerrillos Rd. Santa Fe, NM 87504-1149 Phone: 505 795-3245
New York	Terry Smith Environmental Analysis Bureau New York State Department of Transportation State Campus, 5-303 Albany, New York 12232 Phone: 518-457-2385 Email: tcsmith@dot.ny.us	Gary Frederick Materials New York State Dept. of Transportation 50 Wolf Road, Mail POD 34 Albany, New York 12232 Phone: 518-457-4645
North Carolina	Gregory A. Smith North Carolina Department of Transportation Traffic Noise & Air Quality Group Leader Human Environment Unit 1598 Mail Service Center Raleigh, North Carolina 27699-1598 Phone: 919-431-2010 Email: gasmith@ncdot.gov	Chris Peoples State Materials Engineer North Carolina Department of Transportation P.O. Box 25201 Raleigh, North Carolina 27611-5201 Phone: 919-433-7411
North Dakota	Sheri G. Lares North Dakota Department of Transportation 608 E. Boulevard Avenue Bismark, North Dakota 58505-0700 Phone: 701-328-2555 Email: slares@nd.gov	Terry Woehl Pavement Engineer North Dakota Department of Transportation 300 Airport Road Bismarck, North Dakota 58504 Phone: 701-328-3521

Local Agency	Noise Engineer Contact	Materials / Pavement Engineer Contact
Ohio	Noel Alcala Ohio Department of Transportation Office of Environmental Services 1980 West Broad Street Columbus, Ohio 43223 Phone: 614-466-5222 Email: Noel.Alcala@dot.state.oh.us	Lloyd Welker Materials Engineer Ohio Dept. of Transportation 1980 W. Broad Street Columbus, Ohio 43223 Phone: 614-275-1351
Oklahoma	Kevin Larios Oklahoma Department of Transportation 200 Northeast 18th Street Oklahoma City, Oklahoma 73105 Phone: 405-522-4420 Email: klarios@odot.org	Jeff Dean Pavement/PM Engineer Oklahoma Department of Transportation 200 NE 21st Street Oklahoma City, Oklahoma 73105 Phone: 405-522-0988
Oregon	Carole Newvine Environmental Services Oregon Department of Transportation 355 Capitol Street, NE Room 301 Salem, Oregon 97301 Phone: 503-986-3447 Email: Carole.Newvine@odot.state.or.us	Jeff Gower State Construction and Materials Engineer Oregon Dept of Transportation 411 Trans. Bldg., Room. 200 Salem, Oregon 97310 Phone: 503-986-3123
Pennsylvania	Danielle Shellenberger Pennsylvania Department of Transportation Bureau of Design P.O. Box 3790 Harrisburg, Pennsylvania 17120 Phone: 717-783-6503 Email: dashellenb@state.pa.us	Tim Ramirez Materials Engineer Pennsylvania Department of Transportation 400 North Street Floor 6 Harrisburg, Pennsylvania 17120-0041 Phone: 717-783-6602
Puerto Rico	Luis Rodriquez Puerto Rico Highway and Transportation Authority Post Office Box 42007 San Juan, Puerto Rico 00940 Phone: 787-721-8787 Email: lrodriquez@act.dtop.gov.pr	Orlando Diaz-Quirindongo Materials Engineer Puerto Rico Highway and Transportation Authority P. O. Box 42007 San Juan, Puerto Rico 00940-2007 Phone: 787-729-1592
Rhode Island	Pamela Springer Road Engineering Rhode Island Department of Transportation 2 Capitol Hill, Rm. 234 Providence, Rhode Island 02903 Phone: 401-222-2023 x4660 Email: pspringer@dot.ri.gov	Mark Felag, P.E. Managing Engineer, Materials Rhode Island Department of Transportation 2 Capitol Hill RIDOT Materials Section Providence, Rhode Island 02903 Phone: 401-222-2524 x4130

Local Agency	Noise Engineer Contact	Materials / Pavement Engineer Contact
South Carolina	Randy Williamson Environmental Management Office South Carolina Department of Transportation Post Office Box 191 Columbia, South Carolina 29201 Phone: 803-737-1861 Email: williamsrd@scdot.org	Merrill Zwanka State Materials Engineer South Carolina Department of Transportation P.O. Box 191 Columbus, South Carolina 29202-0191 Phone: 803-737-6694
South Dakota	Alice Whitebird South Dakota Department of Transportation 700 E. Broadway Avenue Pierre, South Dakota 57501-2586 Phone: 605-773-3309 Email: Alice.Whitebird@state.sd.us	Tom Grannes Materials Engineer South Dakota Department of Transportation 700 East Broadway Avenue Pierre, South Dakota 57501-3567 Phone: 605-773-3428
Tennessee	Jim Ozment Tennessee Department of Transportation Office of Environmental Planning and Permits Suite 900, James K. Polk Building 505 Deaderick Street Nashville, Tennessee 37243-0334 Phone: 615-741-5373 Email: jim.ozment@state.tn.us	Gary Head Materials/Testing Director Tennessee Dept. of Transportation James K. Polk Building 505 Deaderick Street, Suite 700 Nashville, Tennessee 37243-0349 Phone: 615-350-4101
Texas	Ray Umscheid Texas Department of Transportation Environmental Affairs Division 125 E. 11th Street Austin, Texas 78701-2409 Phone: 512-416-3025 Email: ray.umscheid@txdot.gov	Jeff Seiders Pavement & Materials Engineer Texas Department of Transportation 125 East 11th Street Austin, Texas 78701 Phone: 512-506-5808
Utah	Stan Adams Utah Department of Transportation Environmental Division Box 148450 Salt Lake City, Utah 84114-8450 Phone: 801-965-4035 Email: stanadams@utah.gov	Gary Kuhl Pavement Engineer Utah Department of Transportation 4501 South 2700 West Salt Lake City, Utah 84119 Phone: 801-964-4552
Vermont	Jeff Ramsey Program Development Division Vermont Agency of Transportation National Life Building, Drawer 33 Montpelier, Vermont 05633-5001 Phone: 802-828-1278 Email: jeff.ramsey@state.vt.us	Mark Woolaver Paving Engineer Vermont Agency of Transportation State Administration Building 133 State Street Montpelier, Vermont 05634 Phone: 802-828-1475

Local Agency	Noise Engineer Contact	Materials / Pavement Engineer Contact
Virginia	Paul Kohler VDOT - Environmental Division 1401 East Broad Street Richmond, Virginia 23219 Phone: 804-371-6766 Email: paul.kohler@vdot.virginia.gov	Charles (Andy) Babish State Materials Engineer Virginia Department of Transportation 6000 Elko Tract Rd Sandston, Virginia 23150 Phone: 804-328-3102
Washington	Tim Sexton Washington State Department of Transportation P.O. Box 330310, NB82-138 Seattle, Washington 98133-9710 Phone: 206-440-4549 Email: sextonT@wsdot.wa.gov	Jeff Uhlmeyer State Pavement Engineer Washington State Department of Transportation Transportation Building P.O. Box 47365 Olympia, Washington 98504-9365 Phone: 360-709-5485
West Virginia	Lovell Facemire West Virginia Department of Transportation State Capitol Complex Building 5, Room A-450 1900 Kanawha Boulevard, East Charleston, West Virginia 25305-0430 Phone: 304-558-2885 Email: lovell.r.facemire@wv.gov	Aaron Gillispie Director, Materials Engineer West Virginia Div. of Highways 190 Dry Branch Road Building Five, Room 110 Charleston, West Virginia 25306 Phone: 304-558-3160
Wisconsin	Jay Waldschmidt Wisconsin Department of Transportation Bureau of Equity and Environmental Services Room 451 4802 Sheboygan Avenue Post Office Box 7965 Madison, Wisconsin 53707-7965 Phone: 608-267-9806 Email: jay.waldschmidt@dot.wi.gov	Steve Krebs Quality Management Engineer Wisconsin Department of Transportation Truax Materials Center 3502 Kinsman Blvd. Madison, Wisconsin 53704 Phone: 608-246-7930
Wyoming	Timothy M. Carroll, P.E. Wyoming Department of Transportation 5300 Bishop Boulevard Cheyenne, Wyoming 82009-3340 Phone: 307-777-4378 Email: timothy.carroll@dot.state.wy.us	Rick Harvey State Materials Engineer Wyoming Department of Transportation P.O. Box 1708 Cheyenne, Wyoming 82003-1708 Phone: 307-777-4476

REFERENCES

General, tire-pavement noise

Rasmussen 2007 — Rasmussen, Robert O. et al., *The Little Book of Quieter Pavements*, Report No. FHWA-IF-08-004, U.S. Department of Transportation Federal Highway Administration (2007).

Sandberg 2002 — Sandberg, Ulf and Jerzy Ejsmont, *Tyre/Road Noise Reference Book*, INFORMEX Ejsmont & Sandberg Handelsbolag, Kisa, Sweden (2002).

Measurement Methodology

Anderson 2004 — Anderson, Grant S., Cynthia S.Y. Lee, Gregg G. Fleming, and Christopher W. Menge, *FHWA Traffic Noise Model, Version 1.0: User's Guide*, Report No.s FHWA-PD-96-009 and DOT-VNTSC-FHWA-98-1, U.S. Department of Transportation, John A. Volpe National Transportation Systems Center, Massachusetts (1998, TNM v2.5 Addendum 2004).

AASHTO CTIM — *Standard Method of Test for Determining the Influence of Road Surfaces on Traffic Noise Using the Continuous-Flow Traffic Time-Integrated Method (CTIM)*, American Association of State Highway and Transportation Officials, AASHTO Specification TP 99 (2012).

AASHTO OBSI — *Standard Method of Test for Measurement of Tire/Pavement Noise using the On-Board Sound Intensity (OBSI) Method*, American Association of State Highway and Transportation Officials, AASHTO Specification TP 76 (2012).

AASHTO SIP — *Standard Method of Test for Determining the Influence of Road Surfaces on Vehicle Noise Using the Statistical Isolated Pass-By Method (SIP)*, American Association of State Highway and Transportation Officials, AASHTO Specification TP 98 (2012).

Donavan 2009 — Donavan, P and Lodico D., *Measuring Tire-Pavement Noise at the Source*" National Cooperative Highway Research Program Report 630 (2009).

Fleming 1996 — Fleming, Gregg G., Amanda S. Rapoza, and Cynthia S.Y. Lee, *Development of National Reference Energy Mean Emission Levels for the FHWA Traffic Noise Model (FHWA TNM®), Version 1.0*, Report Nos. FHWA-PD-96-008 and DOT-VNTSC-FHWA-96-2, U.S. Department of Transportation, John A. Volpe National Transportation Systems Center, Massachusetts (1996).

ISO 11819-1 — *Acoustics - Method for Measuring the Influence of Road Surfaces on Traffic Noise – Part 1: The Statistical Pass-By Method*, International Standard ISO 11819-1, International Organization for Standardization, Geneva, Switzerland (1997).

ISO/CD 11819-2 — *Acoustics - Method for Measuring the Influence of Road Surfaces on Traffic Noise – Part 2: The Close-Proximity Method*, Draft International Standard ISO 11819-2, International Organization for Standardization, Geneva, Switzerland (2010).

Lee 1996 — Lee, Cynthia and Fleming, Gregg, *Measurement of Highway-Related Noise*, Report FHWA-PD-96-046, U.S. Department of Transportation, John A. Volpe National Transportation Systems Center, Massachusetts (1996).

Rochat 2009-1 — Rochat, Judith L., *Developing U.S. wayside methods for measuring the influence of road surfaces on traffic noise*, Proceedings of Inter-Noise 2009, Ottawa, Canada (2009).

FHWA

FHWA Web-1 — Highway traffic noise Web page, accessed December 30, 2010, http://www.fhwa.dot.gov/environment/noise/.

References **Quieter Pavements Guidance Document**

FHWA Web-2	Tire-pavement noise Web page, accessed December 30, 2010, http://www.fhwa.dot.gov/environment/noise/tire_pavement_noise/research/.
FHWA Web-3	Traffic Noise Model® (TNM®) Web page, accessed December 30, 2010, http://www.fhwa.dot.gov/environment/noise/traffic_noise_model/.

Arizona

ADOT 2006	*Progress Report No. 2, Quiet Pavement Pilot Program*, Arizona Department of Transportation (revised December 28, 2006).
ADOT 2008	*Standard Specifications for Road and Bridge Construction*, Arizona Department of Transportation (2008). Related information can be found at http://www.azdot.gov/Highways/ConstGrp/contractors/PDF/2008StandardSpecifications.pdf.
ADOT Web	ADOT's Quiet Roads Web site, accessed December 30, 2010, http://www.azdot.gov/quietroads/index.asp.
AASHTO CTIM	*Standard Method of Test for Determining the Influence of Road Surfaces on Traffic Noise Using the Continuous-Flow Traffic Time-Integrated Method (CTIM)*, Federal Highway Administration Technical Working Group on Tire-Pavement Noise, methodology submitted to AASHTO for consideration as provisional standard (2010).
AASHTO SIP	*Standard Method of Test for Determining the Influence of Road Surfaces on Vehicle Noise Using the Statistical Isolated Pass-By Method (SIP)*, Federal Highway Administration Technical Working Group on Tire-Pavement Noise, methodology submitted to AASHTO for consideration as provisional standard (2010).
Brown 2008	Brown, Vi, *Final Report 584, Survey of Traffic Noise Reduction Products, Materials, and Technologies*, Arizona Department of Transportation (December 2008).
Rochat 2009-1	Rochat, Judith L., *Developing U.S. wayside methods for measuring the influence of road surfaces on traffic noise*, Proceedings of Inter-Noise 2009, Ottawa, Canada (2009).
Scofield 2003	Scofield, L. and Donavan, P., *Development of Arizona's Quiet Pavement Research Program*, Asphalt Rubber Conference, Brasilla, Brazil (December 2003).

California

Caltrans 2005	*I-80 Davis OGAC Pavement Noise Study, 7th Year Summary Report*, California Department of Transportation (December 22, 2005).
Caltrans 2009	*PPB 09-02 Quieter Pavement Strategies for Noise Sensitive Areas*, California Department of Transportation (October 15, 2009). Related information can be found at http://www.dot.ca.gov/hq/esc/Translab/ope/QP-Memo-&-Bulletin-Oct-15-09.pdf.
Caltrans 2010	*Caltrans Thin Lift Study: Effects of Asphalt Pavements on Wayside Noise*, Caltrans Report No. CA 10-0146 (September 2010).
Caltrans Web	Caltrans' Quieter Pavements Web site, accessed December 30, 2010, http://www.dot.ca.gov/hq/esc/Translab/ope/QuieterPavements.html
Lu 2009	Lu, Qing et al., *Investigation of Noise and Durability Performance Trends for Asphaltic Pavement Surface Types: Three-Year Results*, University of California Pavement Research Center, Research Report UCPRC-RR-2009-01 (January 2009).

Pommerenck 2009   Pommerenck, K. and Donavan, P., *Results from the Interstate 5 at Florin Road Noise Monitoring of Open Graded Friction Course with Rubberized Asphalt Concrete (RAC (O))*, prepared for the California Department of Transportation, North Region Environmental Planning, 703 B Street, Marysville, CA 95901 (February 2009).

Rochat 2009-2   Rochat, Judith L. and David R. Read, "Noise Benefits of Asphalt Pavements – Trends at Ages up to 52 Months," Noise Control Engineering Journal, Volume 57 (2) (March-April 2009).

Colorado

CDOT 2005   *Standard Specifications for Road and Bridge Construction*, Colorado Department of Transportation (2005). Related information can be found at http://www.coloradodot.info/business/designsupport/construction-specifications/2005-construction-specs/2005book.

CDOT Brochure   *Highway Traffic Noise: Effect of Pavement Types*, Colorado Department of Transportation, Brochure.

Hanson 2004-1   Hanson, D. and R. James, *Colorado DOT Tire-pavement Noise Study*, Colorado Department of Transportation, Report No. CDOT-DTD-R-2004-5 (April 2004).

Hanson 2006   Hanson, D. and B. Waller, *2005 Colorado DOT Tire-pavement Noise Study*, Colorado Department of Transportation, Report No. CDOT-2006-18, Final Report (November 2006).

Rasmussen 2009   Rasmussen, R. and R. Whirledge, *Tire-pavement and Environmental Traffic Noise Study*, Colorado Department of Transportation, Report No. CDOT-2009-6, Interim Report – 2007 Testing, (June 2009).

Florida

FDOT 2010   *Standard Specifications for Road and Bridge Construction*, Florida Department of Transportation (2010). Related information can be found at http://www.dot.state.fl.us/specificationsoffice/Implemented/SpecBooks/2010BK.shtm.

Wayson 2009   Wayson, R., J. MacDonald, and A. Martin, *Pavement Noise Research, Modeling of Quieter Pavements in Florida*, Florida Department of Transportation, FDOT Project No. #BD550/RPWO#09, Final Report (October 2009).

Kansas

Brennan 2006   Brennan, J. and G. Schieber, *US-69 Surface Texture Noise Study*, Kansas Department of Transportation, Report No. KS-05-3, Final Report (February 2006).

KDOT 2007   *Standard Specifications for State Road and Bridge Construction*, Kansas Department of Transportation (2007). Related information can be found at http://www.ksdot.org/burConsMain/specprov/2007SSDefault.asp.

New Jersey

Bennet 2004   Bennet, T., D. Hanson, and A. Maher, *Demonstration Project – The Measurement of Pavement Noise on New Jersey Pavements Using the NCAT Noise Trailer*, New Jersey Department of Transportation, Report No. FHWA-NJ-2003-021, Final Report (May 2004).

NJDOT 2007   *Standard Specifications for Road and Bridge Construction*, New Jersey Department of Transportation (2007). Related information can be found at http://www.state.nj.us/transportation/eng/specs/2007/Division.shtml.

References **Quieter Pavements Guidance Document**

Ohio

ODOT 2005-1 *Supplemental Specification 803, Rubberized Open Graded Asphalt Friction Course*, Ohio Department of Transportation (April 15, 2005). Related information can be found at http://www.dot.state.oh.us/Divisions/ConstructionMgt/Specs%20and%20Notes%20for%202005/803_04152005%20for%202005.PDF.

ODOT 2005-2 *Effectiveness of Tire/Road Noise Abatement through Surface Retexturing by Diamond Grinding for Project SUM-76-15.40*, Ohio Department of Transportation, Report No. FHWA/OH-2005/009, Final Report (June 2005).

ODOT 2010-1 *Construction and Material Specifications*, Ohio Department of Transportation (January 1, 2010).

ODOT 2010-2 *Pavement Design Manual*, Ohio Department of Transportation (2010).

Texas

AASHTO SIP *Standard Method of Test for Determining the Influence of Road Surfaces on Vehicle Noise Using the Statistical Isolated Pass-By Method (SIP)*, Federal Highway Administration Technical Working Group on Tire-Pavement Noise, methodology submitted to AASHTO for consideration as provisional standard (2010).

Anderson 2004 Anderson, Grant S., Cynthia S.Y. Lee, Gregg G. Fleming, and Christopher W. Menge, *FHWA Traffic Noise Model, Version 1.0: User's Guide*, Report No.s FHWA-PD-96-009 and DOT-VNTSC-FHWA-98-1, U.S. Department of Transportation, John A. Volpe National Transportation Systems Center, Massachusetts (1998, TNM v2.5 Addendum 2004).

Trevino 2006 A Research Plan for Measuring Noise Levels in Highway Pavements in Texas, Texas Department of Transportation, Report No. FHWA/TX-07/0-5185-1, July 2006 (Revised November 2006).

Trevino 2009-1 Trevino-Frias, M. and T. Dossey, *Noise Measurements of Highway Pavements in Texas*, Texas Department of Transportation, Report No. FHWA/TX-10/0-5185-3, April 2009 (Revised October 2009).

Trevino 2009-2 Trevino-Frias, M. and T. Dossey, *On-Board Sound Intensity Testing of PFC Pavements in Texas*, Noise Control Engineering Journal, Volume 57 (2) (March-April 2009).

TxDOT 2004 *Standard Specifications for Construction and Maintenance of Highways, Streets, and Bridges*, Texas Department of Transportation (June 1, 2004). Related information can be found at ftp://ftp.dot.state.tx.us/pub/txdot-info/cmd/cserve/specs/2004/standard/s342.pdf.

Virginia

McGhee 2009 McGhee, K., T. Clark, and C. Hemp, *A Functionally Optimized Hot-Mix Asphalt Wearing Course: Part I: Preliminary Results*, Virginia Department of Transportation, Report No. VTRC 09-R20 (April 2009). Related information can be found at http://www.virginiadot.org/vtrc/main/online_reports/pdf/09-r20.pdf.

McGhee 2010 McGhee, K., *A Functionally Optimized Wearing Course*, Presented at Pavement Evaluation 2010, Roanoke, VA (2010).

Washington State

WSDOT 2005 *Quieter Pavements: Options and Challenges for Washington State*, Washington State Department of Transportation (May 2005).

References **Quieter Pavements Guidance Document**

WSDOT 2009   *Quieter Pavement Tests: The Story So Far*, Washington State Department of Transportation, Brochure (February 2009).

WSDOT 2010   *Standard Specifications for Road, Bridge, and Municipal Construction*, Washington State Department of Transportation, M 41-10, (2010). Related information can be found at http://www.wsdot.wa.gov/publications/manuals/fulltext/M41-10/SS2010.pdf.

WSDOT Web   WSDOT Quieter Pavement Evaluation Web site, accessed December 30, 2010, http://www.wsdot.wa.gov/Business/materialslab/quieterpavement.

Sexton 2010   Sexton, T., *Rapid Deterioration of Sound Level Benefits for "Quieter Pavements" in Washington State Based on the On-Board Sound Intensity (OBSI) Method*, presented at the 159th ASA meeting/Noise-CON 2010 (April 19, 2010).

National Center for Asphalt Technology (NCAT)

Fortier Smit 2007   Fortier Smit, A. and B. Waller, *Sound Pressure and Intensity Evaluations of Low Noise Pavement Structures with Open Graded Asphalt Mixtures*, National Center for Asphalt Technology, NCAT Report 07-02 (June 2007).

Fortier Smit 2008   Fortier Smit, André de, *Synthesis of NCAT Low-Noise HMA Studies*, National Center for Asphalt Technology, NCAT Report 08-01 (March 2008).

Hanson 2004-2   Hanson, D., R. James, and C. NeSmith, *Tire-pavement Noise Study*, National Center for Asphalt Technology, NCAT Report 04-02 (August 2004).

National Concrete Pavement Technology Center (CP Tech Center)

Rasmussen 2008   Rasmussen, R. et al., *How to Reduce Tire-Pavement Noise: Interim Better Practices for Constructing and Texturing Concrete Pavement Surface*, National Concrete Pavement Technology Center (July 2008).

MnROAD

Izevbekhai 2007   Izevbekhai, B., *Report of Diamond Grinding on Cells 7 and 8 MnROAD Mainline Interstate Highway I-94*, Minnesota Department of Transportation (November 2007).

MnROAD 2009-1   *PCC Pavement Surface Characteristics (Rehab) MnROAD Studies TPF 5-(134)*, Minnesota Department of Transportation, MnROAD Brochure (2009).

MnROAD 2009-2   *Pavement Noise Research*, Minnesota Department of Transportation, MnROAD Brochure Version 1 (June 2009).

Directories

Noise Contacts   http://www.adc40.org/adc40membershiplist.pdf.

Pavement Contacts   http://www.fhwa.dot.gov/pavement/pavecont.cfm.

APPENDIX A. BASICS OF TIRE-PAVEMENT NOISE

A.1 Sound vs. Noise

Noise control engineers make an important and clear distinction between the terms *sound* and *noise*. Sound is what you hear. Physically, sound is small air pressure changes traveling as a wave through the air.

Noise is defined as *unwanted* or *extraneous* sound. Obviously, the former *unwanted* part is a subjective definition. What may be considered beautiful, musical sound by one person may be considered just noise by another. In the context of highways and roads, sound from traffic may be considered noise when it becomes too loud or is extraneous to the environment.

A.2 An Analogy

Sound waves are not visible, but they behave similar to waves in water. Imagine standing on the edge of a pond with a calm, smooth surface. You toss a stone in, and circular waves propagate outward from the center of the disturbance. Eventually, the waves reach the edge of the pond where you are standing. Using this analogy, we can explain important concepts about sound.

A.3 Source-Path-Receiver

Using the pond analogy we can define these terms.
1) **Source** – All sounds have a source or generating mechanism. In the case of the pond, the source of the waves is the impact of the stone on the water surface.
2) **Path** – Sound travels through the air from the source to the receiver. In the case of the pond, the path is from the spot where the stone landed to your position on the edge of the pond.
3) **Receiver** – This is the location where the sound is observed by a listener.

Using a quieter pavement to reduce tire-pavement noise is an example of reducing noise at the source. A common method to reduce highway noise is through the use of noise barriers, including earth berms, which reduce noise by affecting the path.

A.4 Units of Sound: dB

Continuing to use the pond analogy, the waves in the pond have crests and troughs. The height of the wave crest above the undisturbed water level indicates the *amplitude* of the wave. Similarly, sound pressure waves in air have pressure peaks and troughs. Pressure amplitude of sound waves is measured in units of Pascals (abbreviated, Pa). All other factors being equal, the greater the pressure amplitude, the louder the sound will be. The lowest pressure amplitude that is audible to people with exceptional hearing is 0.00002 Pa, everyday sounds are in the range 0.002 to 2 Pa, and the threshold of pain is around 100 to 200 Pa. In fact, the range of pressures that is audible is so wide, that a logarithmic scale, the decibel scale, is adopted for sound. The result from this logarithmic transformation is a sound level reported in units of decibels (abbreviated, dB). Figure 4 illustrates the decibel scale along with the corresponding pressure scale for typical sound sources.

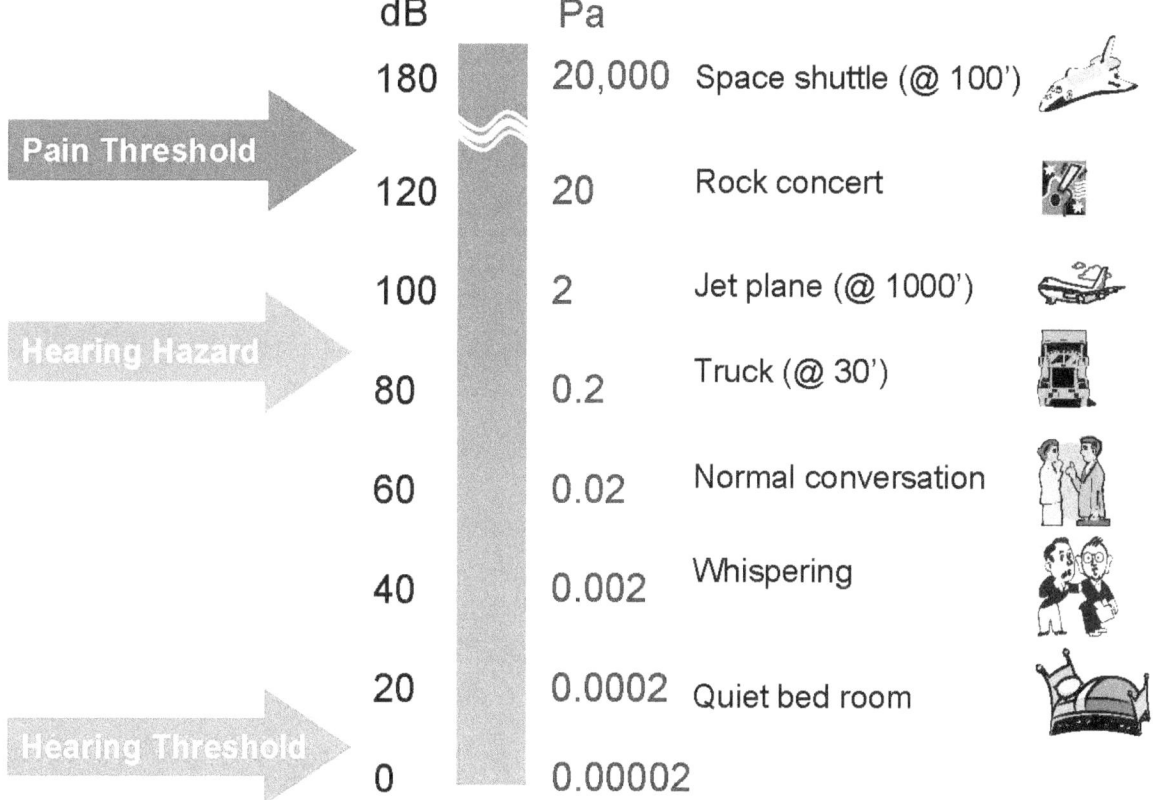

Figure 4. Sound pressure amplitudes and the decibel scale for sound level.

Generally, changes in sound level less than 3 dB are not easily noticed. Changes in sound level of 5 dB are easily noticed. Changes in sound level of 10 dB are perceived as doubling (or halving) of the loudness.

A.5　Frequency and A-Weighting

Again, using the pond analogy from above, the rate at which the wave peaks and troughs arrive at the edge of the pond is called the *frequency* of the wave. Units for frequency are Hertz, abbreviated Hz. For sound waves in air, the frequency of the wave is perceived as pitch.

The frequency range considered audible to the human ear is 20 to 20000 Hz. However, our ears are not equally sensitive to all frequencies. Our ears are most sensitive to frequencies in the range 2000 to 4000 Hz. They are less sensitive to very low and very high frequencies. For this reason, sound is often measured with an A-weighting filter. The A-weighting emphasizes and de-emphasizes frequencies in a manner similar to our ears. (See Figure 5.) Sound levels measured with A-weighting have the letter "A" appended; dBA or dB(A). In the area of highway noise and tire-pavement noise, A-weighting is almost always used, and sound levels are almost always reported in dBA.

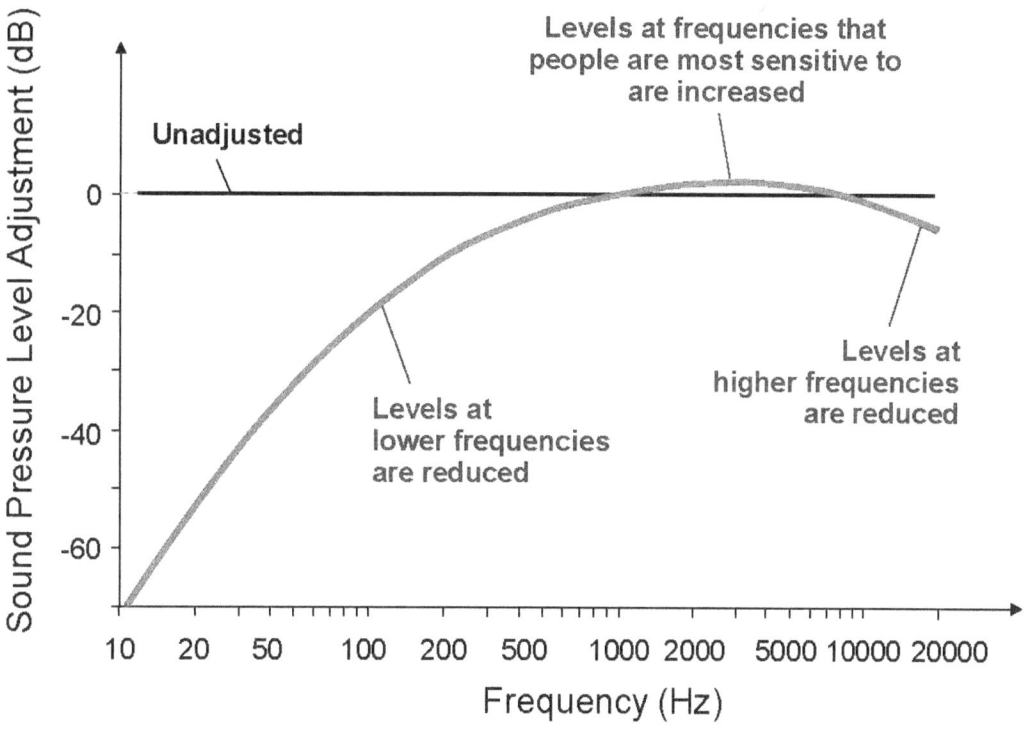

Figure 5. A-weighting curve.

A.6 Traffic Noise and Tire-Pavement Noise

There are many sources of sound on a vehicle driving on a road. These sources are classified into three broad categories.

1) **Propulsion** – This is sound generated by vehicles' engine, exhaust, intake, and other powertrain components.

2) **Aerodynamic** – This is sound caused by air flowing under, over, and around the vehicle.

3) **Tire-pavement** – This is sound generated by the tires rolling on the pavement.

Traffic noise is the total sound generated by all sources from the vehicles traveling on a road. *Tire-pavement* noise is that part of the traffic noise that is due to the tires rolling on the pavement.

When a vehicle starts, stops, and drives at slower speeds, the noise is dominated by propulsion sources, and tire-pavement noise is less significant. As a vehicle speed increases, the tire-pavement noise increases. There is a speed, called the *crossover* speed, above which the tire-pavement noise becomes more significant than the propulsion noise. At speeds greater than the crossover, the tire-pavement noise dominates. Aerodynamic and flow noises do not become important until extremely high speeds. Table 5 lists typical crossover speeds.

Table 5. Typical crossover speeds.

Vehicle Type	Crossover Speed	
	Cruising	Accelerating
Cars	10 – 25 mph (16 – 40 km/h)	20 – 30 mph (32 – 48 km/h)
Trucks	35 – 50 mph (56 – 80 km/h)	> 50 mph (> 80 km/h)

Whenever there is a situation with excessive traffic noise, it is important to understand the contribution of tire-pavement noise to the total traffic noise. Quieter pavements will not reduce traffic noise when the tire-pavement noise is not a significant contributor. For example, if the traffic noise is mostly engine and exhaust noise from cars and trucks slowing down and accelerating at an intersection, a quieter pavement will provide little to no noise reduction. Quieter pavements offer benefit in situations where the traffic is higher speed and free flowing.

A.7   Tire-Pavement Noise Generation

Figure 6 illustrates the typical components of a tire and tread. The portion of the tread in contact with the pavement is referred to as the *contact patch*. When a tire rolls, the tread blocks come into contact with the pavement at the leading edge of the contact patch, and then pull off the pavement at the trailing edge. At higher speeds, the action of the tread blocks coming in and out of contact with the pavement becomes analogous to a series of rubber hammers rapidly impacting the road. The forces, vibration, and airflow occurring during this rolling action generate noise. Grooves and holes in the tread pattern create acoustic channels and cavities that further generate and amplify noise. Noise is also amplified by tire resonances and vibrations of the tire sidewall. When all is considered, the ways a tire and pavement interact to generate noise are many and complex.

Appendix A Quieter Pavements Guidance Document

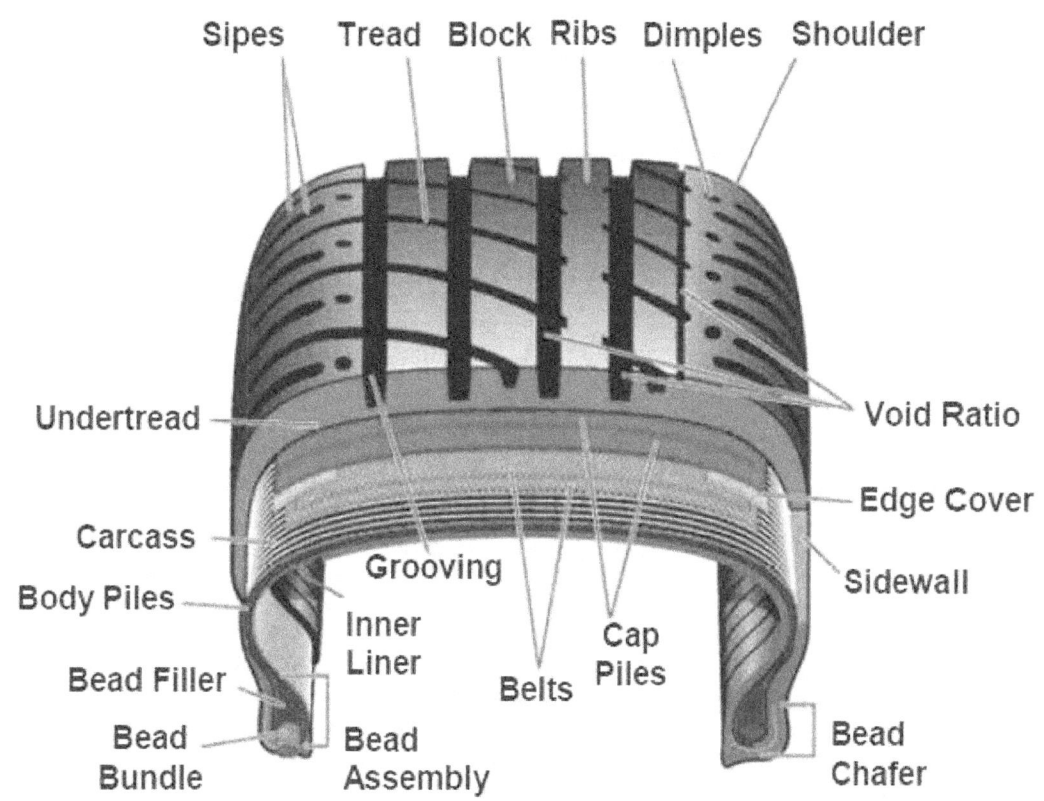

Figure 6. Components of a tire and tread. (Source: Yokohama tires.)

A.8 How Tire-Pavement Noise is Measured

There are two approaches to measuring tire-pavement noise:

1) Wayside (or roadside) methods; and
2) Source methods.

Wayside methods measure noise along the side of the road using microphones positioned at standard distances from the center of the travel lane, often at 25 or 50 ft (7.5 and 15 m). (See Figure 7.) Microphones may also be positioned where human activity occurs; for example, in a residential backyard or park campground. There are several wayside methods available. In situations with little or light traffic, statistical pass-by (SPB), controlled pass-by (CPB), or statistical isolated pass-by (SIP) methods are used. These methods measure the sound levels from individual vehicles traveling on the road. For situations with heavy, continuous flowing traffic, the continuous-flow traffic time-integrated method (CTIM) should be used instead.

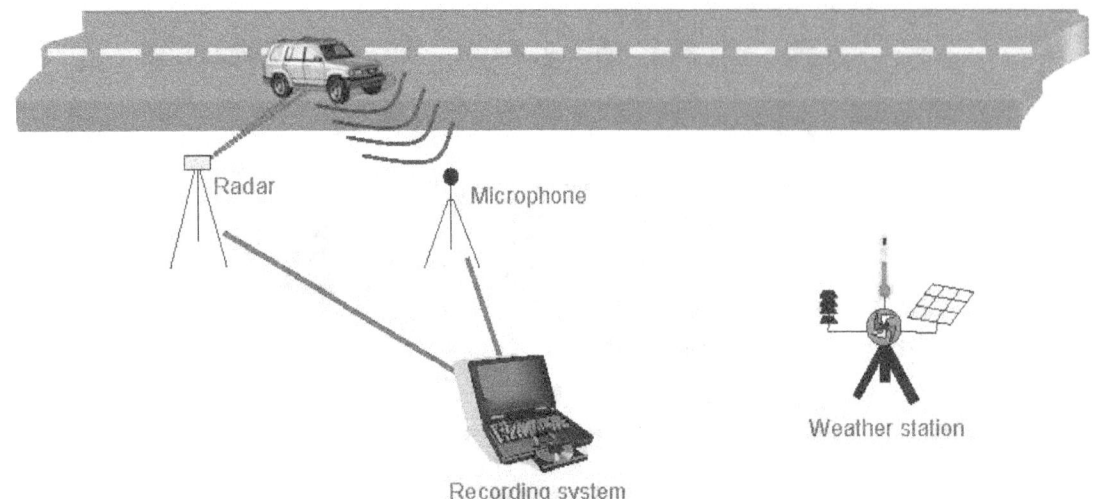
Figure 7. Test setup for wayside noise measurements.

Source methods measure noise near the tire using microphones positioned very close to the tire-pavement interface. There are two standardized source methods: the close proximity (CPX) method and the on-board sound intensity (OBSI) method. The CPX method is often referenced in older research studies and/or those from outside the US. The OBSI method is the accepted method in the USA, and is referenced in most of the quieter pavement research studies. OBSI uses special microphone probes that have directional characteristics favorable for this type of measurement. (See Figure 8.) When conducting tire-pavement noise using any source method, a standard tire is used, which today is the ASTM standard reference test tire (SRTT). A standard vehicle speed is also used since the sound level is a function of this factor.

Figure 8. Photo of microphones in OBSI test configuration.

The Department of the Interior protects and manages the nation's natural resources and cultural heritage; provides scientific and other information about those resources; and honors its special responsibilities to American Indians, Alaska Natives, and affiliated Island Communities.

NPS 999/121284, June 2013

www.ingramcontent.com/pod-product-compliance
Lightning Source LLC
Chambersburg PA
CBHW081855170526
45167CB00007B/3025